3rd Italian-Pakistani Workshop on Relativistic Astrophysics

(IPWRA2011)

Journal of Physics: Conference Series Volume 354

Lecce, Italy
20-22 June 2011

Editors:

F. De Paolis
Azad A. Siddiqui

ISBN: 978-1-62276-771-7
ISSN: 1742-6588

Printed from e-media with permission by:

Curran Associates, Inc.
57 Morehouse Lane
Red Hook, NY 12571

Some format issues inherent in the e-media version may also appear in this print version.

Printed by Curran Associates, Inc. (2013)

Institute of Physics
Dirac House, Temple Back
Bristol BS1 6BE UK

Phone: 44 1 17 929 7481
Fax: 44 1 17 920 0979

techtracking@iop.org

Additional copies of this publication are available from:

Curran Associates, Inc.
57 Morehouse Lane
Red Hook, NY 12571 USA
Phone: 845-758-0400
Fax: 845-758-2634
Email: curran@proceedings.com
Web: www.proceedings.com

TABLE OF CONTENTS

Editorial of the Proceedings of the Third Italian-Pakistani Workshop on Relativistic Astrophysics

F. De Paolis

Dipartimento di Fisica, Università del Salento, Via per Arnesano, I-73100, Lecce, Italy and INFN, Sezione di Lecce, Via per Arnesano, I-73100, Lecce, Italy

E-mail: francesco.depaolis@le.infn.it

Azad A. Siddiqui

Centre for Advanced Mathematics and Physics, National University of Sciences and Technology (NUST), H-12, Islamabad, Pakistan

E-mail: azad@camp.nust.edu.pk

The Third Italian-Pakistani Workshop on Relativistic Astrophysics was held at the Rectorate of the University of Salento in Lecce on June 20-22, 2011. It follows the first two editions of this Workshop held at the Department of Physics of the University of Salento on 20-22 June 2007 and at ICRA (International Center for Relativistic Astrophysics) in Pescara on 8-10 July 2009. The Proceedings of the first two editions of this Workshop have been published on two special issues of Nuovo Cimento B [1] and General Relativity and Gravitation [2], respectively.

The workshop series, whose aim is that of discussing the different aspects (both theoretical and observational) of Relativistic Astrophysics, follows the signature, in 2006, of an agreement between the University of Salento, Italy and the National University of Sciences and Technology (NUST), Pakistan, and aims at promoting scientific and academic cooperation between the parties.

The organizing committee of this Workshop has decided to dedicate the present workshop's edition to the celebration of the 65th birthday of the founder of this series of meetings, Prof. Asghar Qadir, one of the greatest Pakistani scientists of any time and a renowned world expert in the theory of general relativity.

Many of the Workshop's participants have either been students or collaborators of Asghar Qadir, or both. In Pakistan the words Relativity and Asghar Qadir are synonymous. It will not be entirely wrong to say that anybody who has anything to do with relativity in Pakistan is either his student or a student of one of his students. Asghar Qadir has inspired generations of researchers and teachers, and continues to be a source of inspiration for hard work and dedication. He is a mentor of Pakistani scientists and the equivalent in Pakistan of what John Archibald Wheeler has been in the US.

Asghar has the rare privilege of being introduced into the world of research by Sir Roger Penrose, who was his PhD supervisor. They worked on Twistor theory, which was one of the foremost runners in the race to quantize gravity at the time. Later in 1971, he joined Quaid-i-Azam University (then Islamabad University) and developed research collaborations

Figure 1. An autographed picture of John Archibald Wheeler with a young Asghar Qadir

with several well-known physicists and mathematicians including John Archibald Wheeler and Remo Ruffini. Of special mention are his efforts with Abdus Salam (NL) for encouraging Physics and Mathematics in the Third World and specially in Pakistan (by running the Salam Prize in Science). Asghar has since become one of the world's top relativists and a leading name in the study of symmetries. In 2004, he founded the Centre for Advanced Mathematics and Physics at NUST. Under his dynamic leadership, the centre has become as influential as any other institute of Pakistan in the development of Mathematics and Mathematical Physics within a very short time. One of his many talents is his amazing ability to detect potential to do science among his students. As a result he has been able to successfully guide an astonishing number of PhDs, 14 to be exact and still counting, and no doubt this has been his greatest contribution to Relativity in Pakistan. Asghar has received several national and international awards and honours for his services to science in and outside Pakistan.

Figure 2. Celebrations of the 65th birthday of Asghar Qadir during third Italian-Pakistani workshop on Relativistic Astrophysics

These Proceedings contain revised versions of selected papers presented at the Workshop, which has been sponsored by the Salento University, the National University of Sciences and Technology, the Abdus Salam International Center for Theoretical Physics and the National

Institute of Nuclear Physics (INFN, Section of Lecce). All these institutions are acknowledged here for their support together with the bank Monte dei Paschi di Siena.

Figure 3. Participants of the third Italian-Pakistani workshop on Relativistic Astrophysics

We would like to thank all the Workshop's participants, more than 30 physicists coming from at least six countries, and in particular the speakers who contributed with enthusiasm to the scientific success of the Workshop. Special thanks go to the members of the organizing committee, actively involved in the organization of this event, and in particular to Achille Nucita, Gabriele Ingrosso, Mubasher Jamil and Ibrar Hussain.

References

[1] De Paolis F and Qadir A (Eds.) 2007 Nuovo Cimento B **122B** 465-616
[2] De Paolis F, Ellis G F R, Qadir A and Ruffini R (Eds.) 2011 General Relativity and Gravitation **43** 929-1202

Journal of Physics: Conference Series **354** (2012) 012001 doi:10.1088/1742-6596/354/1/012001

Microlensing towards the Magellanic Clouds and M31: is the quest for MACHOs still open?

Sebastiano Calchi Novati

Dipartimento di Fisica "E. R. Caianiello", Università di Salerno, Via Ponte don Melillo, 84084 Fisciano (SA), Italy & Istituto Internazionale per gli Alti Studi Scientifici (IIASS), Vietri Sul Mare (SA), Italy

E-mail: novati@sa.infn.it

Abstract. Microlensing is the tool of choice for the search and the analysis of compact halo objects ("MACHOs"), a still viable class of dark matter candidates at the galactic scale. Different analyses point towards an agreement in excluding dark matter MACHOs of less than about 10^{-1} $M_{\odot}$; it remains however an ongoing debate for values in the mass range $(0.1-1)$ $M_{\odot}$. The more robust constraints, though not all in agreement, come from the observational campaigns towards the Magellanic Clouds (the LMC and the SMC). The analyses towards the nearby galaxy of M31, in the so called "pixel lensing" regime, have expanded the perspectives in this field of research. In this contribution first we draw a critical view on recent results and then we focus on the pixel lensing analysis towards M31 of the PLAN collaboration.

1. Introduction

The original suggestion of Paczyński (1986) of using microlensing as a tool to probe compact halo objects as a dark matter candidate at the galactic scale has given rise, in the meantime, to a much larger range of applications of microlensing which has become an efficient tool of research for many different relevant astrophysical issues. (The observational driving reason being the relatively large rate of events expected, and by now routinely observed, towards the Galactic bulge.) A partial list includes the study of the inner Galactic structure (Moniez, 2010 for a review), many aspects of stellar astrophysics (Gould, 2001) with in particular analyses on the mass function towards the Bulge (Gould, 2000; Calchi Novati et al, 2008) and the recent exciting result on an unbound or distant planetary mass population (Sumi et al, 2011). Currently, however, the main field of application of microlensing has become the search of exoplanets thanks to its ability to explore different part of the planetary parameter space with respect to other techniques (Gaudi, 2010; Dominik, 2010 for recent reviews). Within this framework first results on the frequency of exoplanets have also been reported (Sumi et al, 2010; Gould et al, 2010). Microlensing is indeed well suited to infer the statistical properties of the underlying exoplanet population, still, an adequate strategy to this purpose must be adopted (Dominik et al, 2010).

The search of MACHOs remains however an active field of research, with observational campaigns being carried out to this purpose both towards the Magellanic Clouds and M31. As more thoroughly discussed in the following sections, the debate is currently centered on an hypothetical MACHO population in the mass range $(0.1-1)$ $M_{\odot}$. This is however the very same range of mass for normal stars acting as lenses (we broadly refer to this case as "self lensing", as opposed to MACHO lensing), the lens mass being the driving physical parameter of the observed characteristics of microlensing light curves. This coincidence may therefore be taken as an hint of some bias in the analyses. If not the case, however, the reported sizeable fraction of MACHOs in this mass range (up to about $f = 20\%$ in mass of the halo) would represent a real astrophysical challenge certainly worth more detailed analyses.

2. The results towards the Magellanic Clouds

The main results along this line of sight have been reported by the MACHO, the EROS and the OGLE collaborations. We refer to Moniez (2010) for a detailed discussion on the subject and a full list of references. The relevant and definite result of all these observational campaigns is that compact halo object *are not* a major component of the Galactic halo. However, some disagreement remains to be clarified in some of the considered mass ranges. There is a full consensus to exclude, for a fraction in mass well below 10%, MACHOs as viable dark matter candidates from about 10^{-7} $M_\odot$ up to a mass below about 0.1 $M_\odot$. This agreement then breaks down in the mass range $(0.1 - 1)$ $M_\odot$. For larger values of the MACHO mass the smaller number of expected candidates makes the constraints less tight (although, in a re-analysis of the OGLE-II and OGLE-III data, Calchi Novati and Mancini (2011) obtained the stronger bounds to date by microlensing in this mass range). More in particular, the MACHO collaboration (Alcock et al, 2000a) reported an evidence of a signal, $f \sim 20\%$ (in the hypothesis of a "standard" pseudo-isothermal spherical density distribution for the halo[1]) for about 0.5 $M_\odot$ MACHOs out of observations towards the LMC. Two critical aspects of this analysis are worth being recalled: MACHO monitored mainly the central LMC region, where one can expect a larger contamination by the LMC self-lensing signal; MACHO considered as viable sources also rather faint stars and this complicates the analysis of the detection efficiency and therefore the related aspect of the estimate of the expected signal. The results of the MACHO collaboration, substantially confirmed in a more recent analysis by Bennett (2005), have then been challenged by the EROS collaboration (Tisserand et al, 2007) and more recently by the analyses of the OGLE collaboration whose OGLE-II and OGLE-III campaigns towards both the LMC and the SMC have been discussed in a recent series of papers by Wyrzykowski et al (2009, 2010, 2011a,b). Taken at face value, all these last analyses indicate that self lensing alone is able to explain the observed rate of events, clearly at odds with the result of the MACHO collaboration.

The results of the MACHO collaboration have been the object of several investigations trying to explain the observed rate *without* introducing a dark matter component. Additional (stellar) lens populations contributing to the self-lensing signal (in his more broad exception) may come from local structures along the line of sight (Zaritsky and Lin, 1997; Zhao et al, 2003). Such populations, even if not very massive, may enhance a large enough lensing signal with respect to LMC self lensing because of the larger source-lens distance would imply an increase in the microlensing cross section. For the same reason an enhancement in self lensing would also come from a more complicate LMC density distribution (Zhao and Evans 2000), or from a source population lying *behind* the LMC (Alcock et al, 2001).

Comparing to the case of M31, to be discussed in the next Section, it is worth recalling a few points specific to the observations towards the Magellanic Clouds (in addition to the different microlensing regime one enters when looking at the more distant sources in M31). First, an intrinsic bias in the results may be introduced by the implicit underlying, though not probed, assumption that the single line of sight probed in this case through the Galactic halo is representative of this full component. This may not be the case for instance because of tidal

[1] As it has been already observed, eg Bennett (2005), the fact that MACHOs, if any, are not the main component of dark matter haloes, also implies that they may not follow its spatial distribution. In turn, this implies that the actual mass fraction related to the observed events, if really to be attributed to MACHO lensing, may in fact be smaller than that evaluated for a standard halo model.

interactions with the Galaxy and/or local halo clumpiness. Second, the expected rate by "self lensing" is small (with a sizeable fraction of lenses to be expected from the Milky Way disc, as recently pointed out by Calchi Novati and Mancini, 2011), both in absolute number, at least for the reported observational campaigns, and in a relative sense with respect to MACHO lensing. In fact, the EROS collaboration discussed the MACHO fraction out of a single candidate event observed towards the SMC and none towards the LMC, and the observed rate of the recent OGLE-II and OGLE-III campaigns is of the order of a few events only. This is in fact a bonus as the small "background" signal of self-lensing events with respect to MACHO lensing facilitates the statistical interpretation of the results. However, the small statistics of events at disposal becomes a problem to the extent that it makes difficult an analysis of the event characteristics which may shed more light on the nature of the events beyond the "simple" statistics based essentially on the *number* of events alone (as outlined, eg, in the analyses of Mancini et al, 2004 and Calchi Novati et al, 2006 where the event duration and spatial distributions were used to this purpose). A larger self-lensing statistics would therefore definitively help as it would make self-lensing signal a robust test case so to finally better probe MACHO lensing. The relevance of carrying out more detailed analyses on the lens nature is made apparent also by the thorough analysis of OGLE-2005-SMC-001 (Dong et al, 2007), out of which the authors conclude that "halo lenses are strongly favored but SMC lenses are not definitively ruled out". In fact, Dong et al (2007) conclusions are at odds also with respect to the results of the MACHO collaboration as they favor, to explain the lens, a Milky Way halo $\sim 10\ M_\odot$ *black hole* solution, namely in a mass range already excluded by Alcock et al (2000a). This composite picture strongly motivates the still ongoing observational effort towards the LMC and the SMC carried out by OGLE-IV (Udalski, 2011) and MOA-II (Sumi, 2011).

3. Pixel Lensing towards M31

Lying at about 770 kpc, the nearby giant spiral galaxy of M31 is more than one order of magnitude more distant than the LMC (with a difference in distance modulus between the LMC and M31 of about 6). This makes the potential sources of microlensing events along this line of sight *unresolved* for most ground-based telescope. This is the underlying reason for the different regime one enters in that case, the so-called "pixel lensing" (Gould, 1996). The source being unresolved, the source flux is usually more difficult, if not impossible, to be estimated from the light curve data alone. This complicates the analysis as it adds an additional degeneracy in the microlensing parameter space (M31 is however, fortunately, near enough so that light curve data, if sufficiently well sampled, especially along the wings of the microlensing flux variations, may allow one to break this degeneracy). On the other hand, with respect to the LMC, there is a much larger number of potential sources available within a comparatively much smaller cone of view. (In fact, the sensitivity of past and current observational campaigns allowed to detect only microlensing flux variations out of very bright sources lying at the bright end of the luminosity function, roughly stars brighter than $M_I \sim 2$. Since this is a magnitude range where the luminosity function is still undergoing an extremely steep rise, an increase in sensitivity may therefore bring to hugely extend the number of potential sources and accordingly of the expected signal.) M31 is therefore a target of choice for the search of MACHOs. A main bonus of this line of sight is the possibility to map all of the M31 dark matter halo, which is not possible for the Milky Way one. In fact, at parity of halo mass fraction and MACHO mass, one expects about one third of MACHO events from the Milky Way dark matter halo. This is because the (much) larger number of available lenses in M31 is (almost) exactly compensated by the (much) larger cross section of the Milky Way halo lenses.

The observational results towards M31 have been reviewed in Calchi Novati (2010). We recall the 4 years (1999-2002) observational campaign carried out at the 2.5m INT telescope, whose data have been analysed, in a completely independent way, by the POINT-AGAPE and the MEGA collaborations. Interestingly, the conclusions reached by these two groups on the MACHO issue differ. For POINT-AGAPE, Calchi Novati et al (2005) reported an evidence of a MACHO signal along this line of sight, with in particular a *lower* limit for f of about 20% in the mass range $(0.1-1)$ $M_\odot$. This result has been however challenged by MEGA (de Jong et al, 2006) who concluded that self lensing could explain the full observed rate and placed an *upper* limit on f of about 30% in the same mass range. The small statistics is an issue for the POINT-AGAPE analysis which is based on 5 reported microlensing candidates and for which, in addition, the robustness of the conclusions on the MACHO issue rests mainly upon a single candidate, PA-99-N2, for which, according to the POINT-AGAPE analysis, the large distance from the M31 centre makes very unlikely the attribution to a self-lensing population. The analysis of the MEGA collaboration, based on a more than twice as much larger set of microlensing candidates, has been on the other hand challenged on the basis that the characteristics of the observed events are in fact at odds with a full self-lensing explanation (Ingrosso et al 2007).

Even beyond its importance within the POINT-AGAPE analysis, PA-99-N2 is a very peculiar event. The good coverage and tested achromaticity in three different bands make of it a a very robust candidate in itself. A caveat is the rather large full width at half maximum duration, about 20 days (coupled to a very large flux deviation at maximum), at the very limit of the usually expected shorter values (a prediction confirmed by several other microlensing candidates). The genuine microlensing nature of this flux variation is on the other hand confirmed also by an anomaly along the light curve which turned out to be compatible with a binary lens model (An et al, 2004). Indeed, it was subsequently observed that the secondary lens could be due to a planetary system (Ingrosso et al 2009, 2011) although the large uncertainty on the model parameters does not allow sharp conclusions on this issue. (This would make of PA-99-N2 the second extra-galactic exoplanet *candidate* possibly ever observed after that of the microlensing event MACHO LMC-1, the very first event reported by the MACHO collaboration towards the LMC, Alcock et al 1993, which remarkably showed an anomaly that could even be explained by an exoplanet secondary lens Dominik & Hirshfeld 1994, 1996; Alcock et al, 2000b).

The disagreement of the MEGA and the POINT-AGAPE analyses may be traced back to two main lines of explanation, both turning around the problem of the estimate of the expected signal. First, the evaluation of the expected self lensing signal, due to the M31 luminous populations, still suffers from possibly inaccurate modeling of these components. Second, the correct evaluation of the *efficiency* of a given pipeline, the essential link to meaningfully compare the theoretical prediction with the observed rate is, even more for M31 than for the LMC case, highly non trivial.

To complicate the analysis one has to face the fact that the expected self lensing signal, at least in the central M31 region, is about of the same order of magnitude of an hypothetical MACHO population in the stellar mass range. This complication comes together with the additional degeneracy of the lensing parameter space peculiar to the pixel lensing regime which makes usually difficult reliable estimates on the physical time scale of the event and therefore even more difficult any investigation on the lens *nature*. In fact the statistics of choice to disentangle MACHO lensing from self lensing is the event spatial distribution, with self lensing expected to be, relatively, more clustered around the M31 centre (in fact, whatever the lens population, the signal gets larger towards the M31 centre following the spatial density distribution of the sources, at least as long it is not severely suppressed by the huge background noise of the underlying M31 surface brightness profile). As a second order effect, M31 MACHO lensing, because of the inclination of M31 along the line of sight, is expected to show an asymmetry in the spatial distribution which should help to characterize this signal (Crotts, 1992; Baillon et al 1993; Jetzer 1994). The (recurrent) caveat is the small statistics at disposal which has made difficult, at least up to now, to efficiently carry out this kind of analyses.

Parallel to the statistical analysis of full data set, as those of POINT-AGAPE and MEGA, the importance of carrying out thorough analyses of *single* events has become in the meantime increasingly clear. Whenever they are possible, namely when the data are endowed with a sufficiently good sampling along the flux variation, these analyses are extremely useful as they allow one to extract additional and useful information on the lens nature. This has probed to be the case in particular for two events, both enjoying coverage by two different data set. (Incidentally, the routinely performed merging of several data set for events observed and followed up towards the Galactic bulge should not lead to underestimate the importance of having successfully carried out a similar analysis for M31 pixel lensing events.) The PA-S3/GL1 event, first reported by the POINT-AGAPE collaboration (Paulin-Henriksson et al, 2003) and then by the WeCAPP collaboration (Riffeser et al, 2003) has been subsequently analysed in more detail by Riffeser et al (2008). Based on the extremely large flux deviation at maximum amplification the authors of this last work conclude, even for an event lying in the very inner M31 region, strongly in favor of a MACHO hypothesis for the lens nature. A similar conclusion, even if with a lower statistical significance, was reached in the analysis of the OAB-N2 event carried out by the PLAN collaboration (Calchi Novati et al 2010). In this case the analysis was based on the study of the finite size source effect and the related study of the lens proper motion made possible by the excellent sampling along the full microlensing flux variation together with a careful analysis on the unlensed source flux based on archive HST data.

More recently, Lee et al (2011) presented the first results of a new challenging observational M31 pixel lensing project, PAndromeda, carried out at the 1.8m PS1 telescope with a huge field of view (7 $\deg^2$). In particular they report 6 new candidate events out of their first season of data, for an analysis restricted to a subfield of $40' \times 40'$ around the M31 centre, showing in particular excellent flux stability. New data and analyses from this project are expected.

3.1. The PLAN campaign at OAB[2]

As a PLAN (Pixel Lensing Andromeda) collaboration we have being carrying out a pixel lensing observational campaign towards M31 using the 152cm Cassini telescope located at "Osservatorio Astronomico di Bologna" in central Italy (at declination 44°.4, perfectly suitable to observe M31). Following a shorter pilot season in 2006, during the following 4 years PLAN has been allocated about 50 (almost) consecutive full nights/year.

Table 1. Statistics of the observational sampling for the PLAN campaign at OAB: (1): year of observation; (2): number of good over allocated nights; (3) average number of hours, per good night, of M31 observations.

(1)	(2)	(3)
2006	8/11 (73%)	4.2
2007	31/50 (62%)	3.8
2008	38/65 (58%)	4.6
2009	25/36 (69%)	5.5
2010	20/41 (49%)	4.6
tot	122/203 (60%)	4.5

The observational strategy is set so to match the characteristics of the expected signal and maximise, at the same time, the expected rate. Microlensing variations in M31 are typically expected to last (in term of the full width at half maximum duration, $t_{1/2}$, which characterize

[2] Results reported on behalf of the PLAN collaboration.

Journal of Physics: Conference Series **354** (2012) 012001 doi:10.1088/1742-6596/354/1/012001

the observed signal) a few days only. A regular sampling with high time-resolution along the expected flux variations is therefore essential first to distinguish microlensing variations from other kind of intrinsic variable signals, second to properly characterize the lensing parameter space (as mentioned, a good enough sampling can even allow one to break the degeneracy and get reliable estimate of the physical duration, the Einstein time $t_{\rm E}$). As a result, a long enough baseline (at least 2-3 weeks) of consecutive nights is essential to carry out such an observational programme. A longer baseline is then useful both to increase the statistics of expected events and to better constrain microlensing against other variable signals. Achromaticity is a further peculiar characteristic of microlensing useful to be exploited so that observations should preferentially be carried out in two bands (in particular, the expected sources being typically luminous red giants, we have used two broad-band filters similar to R, I Cousin). To reach a large enough signal-to-noise level is also essential as we have to overcome the large background noise set by the M31 surface brightness profile. With a 1.5m class telescope we have therefore to get to a long enough equivalent integration time per night, at least of the order of 1-2 hours per filter. With a CCD field of view of $13' \times 13'$ (pixel scale of $0.58''$), we have carried out observations towards 2 fields respectively North and South the M31 centre, leaving out the inner 2 arcmin. This way we cover the inner M31 part, in fact almost all of the M31 bulge region, still we reach far enough away from the M31 centre so to be able, at least in principle, to disentangle MACHO lensing from self lensing on the basis of the spatial distribution of the events. Overall our observational strategy, requiring about 8 hours per night (2 hours/field/filter), is set to match the number of potentially available hours per night with our target comfortably below an airmass of 1.5. Full details on the set up and observational strategy are further discussed in Calchi Novati et al (2007).

The statistics of the observations are reported in Table 1. Overall, the average weather conditions (humidity, cloud coverage) have not shown to be optimal to our purposes, introducing a lot of unwanted gaps in the sampling. The ratio of at least partially clear nights has indeed remained at about the 60% level, which is barely sufficient, with however in addition an average number of hours we could actually observe M31 during those nights below 5 hours, so that the overall fraction of hours useful to our purposes, with respect to those allocated and potentially usable, turned out to be in fact well below 40%.

The search for flux variations compatible with a microlensing signal is carried out with a fully *automated* selection pipeline. The result of this analysis are then compared to the expected signal evaluated through a full Monte Carlo simulation of the experiment, where the observational set up is reproduced together with the astrophysical and the microlensing amplification model (in particular, we duly include the finite size source effect). The events *selected* within the Monte Carlo are then simulated on the *raw* data upon which we apply from scratch the selection pipeline (we stress that in this last step the simulation is being carried out on the *images*). This way we get to an extremely reliable estimate of the *efficiency* of the selection pipeline which in turn allows us a correct estimate (for a given astrophysical model) of the number of expected events.

Full details of the selection process, the Monte Carlo simulation and the related efficiency simulation as well as the adopted astrophysical model are given in Calchi Novati et al (2009) where we have also reported the results of the 2007 season of data. In particular, two flux variations passed the full selection pipeline and accordingly have been classified as microlensing *candidates*: OAB-N1 and OAB-N2. Although primarily made to the purpose of maximising the efficiency (and the observed rate), the selection pipeline is also intended to leave the final sample of candidates completely free from contaminations of intrinsic variable sources (the only viable option so to draw some meaningful conclusion when dealing with a signal of *a few* expected events compared to *thousands* of detected flux variations). Still, with few exceptions, the fate of microlensing events is to keep their attribution of *candidates* if not for being altogether rejected.

In this specific case, both detected flux variations had some features that did not make them appealing. As a main problem, OAB-N1 is affected by noisy data, but this could in fact simply reflect the overall quality of our data set, and indeed we could not find any plausible alternative explanation to microlensing (and its estimated duration is well in agreement with the expected signal). OAB-N2 is a high signal-to-noise flux variation nicely following the microlensing shape. However, as presented in Calchi Novati et al (2009), it completely lacks data points along the descent. The genuine microlensing nature of OAB-N2 was to be strongly supported in a second moment when additional data, in particular useful to properly sample the descent, were kindly made available to us by the WeCAPP collaboration (Calchi Novati et al 2010).

The estimate of the expected signal resulted in about 1 self-lensing event, and about the same for *full* haloes (both M31 and the Milky Way) of 0.5 $M_\odot$ MACHOs (and about twice as much for 10^{-2} $M_\odot$ MACHOs). Comparing to the two candidate microlensing events the observed rate looks therefore in agreement with the expected self-lensing rate, the expected MACHO signal on the other hand being too small to allow us to draw meaningful conclusions.

The perspective on this initial tentative result was to be modified by the already reported thorough analysis of OAB-N2 (once the sampling completed by the WeCAPP data) which allowed us to conclude that OAB-N2 should be, though not with very large significance, attributed to MACHO lensing rather than to self lensing (Calchi Novati et al 2010). We are currently completing the analysis of the remaining three years of data (Calchi Novati et al, 2012). Hopefully, the enlarged statistics is going to allow us to draw sharper conclusions on the candidates lens nature and the MACHO issue.

4. Conclusions

Nowadays the effort of the observational microlensing community is mainly driven by the search of exoplanets. The search of dark matter at the galactic scale in the form of compact halo objects remains however an important and challenging open issue. In fact, this second line of research is susceptible to benefit from the exoplanet effort and this opportunity should not be lost. The upgraded OGLE-IV and MOA-II systems, even with their main target being the Galactic bulge, will give the chance to proceed to new and more thorough researches of MACHOs towards the Magellanic Clouds. As for M31, a main challenge is to enlarge the observed rate up to a level for which first, the MACHO issue could be fully addressed on the basis of a large enough statistics. Second, a large enough rate would allow parallel lines of research to be addressed, as those carried out towards the Galactic bulge, first of all the search of exoplanets (lines of research for which microlensing would constitute one of the few, if not the only, available tool). A further challenge is to get, for M31 as already currently done for the Galactic bulge targets, to a survey alert system coupled to a follow up round-the-clock coverage of the ongoing events. To this purpose, however, one has to face the difficulty linked to acknowledge the microlensing nature of a pixel lensing variation already during its raising part, a difficulty enhanced by the usually rather short duration of the flux variations involved. This remains however a compelling step to fully address the issue of the correct understanding of the lens parameter space.

Microlensing observational campaigns have robustly established that compact halo objects *do not* constitute the main fraction of galactic dark matter haloes. The picture is still to be fully understood, however, in particular for MACHO in the mass range $(0.1 - 1)$ $M_\odot$: a sizeable fraction of these objects, as that suggested by some results, would indeed in any case represent a relevant challenge to our understanding of galactic astrophysics. The larger statistics of events expected from the ongoing campaigns towards both the Magellanic Clouds and M31 promise to help to finally solve the puzzle.

References

Alcock C et al 2001, *ApJ* **552**, 582
Alcock C et al 2000a, *ApJ* **542**, 281
Alcock C et al 2000b, *ApJ* **541**, 270
Alcock C et al 1993, *Nature* **365**, 621
An J et al 2004, *ApJ* **601**, 845
Baillon P, Bouquet A, Giraud-Héraud Y and Kaplan J 1993, *A&A* **277**, 1
Bennett D 2005, *ApJ* **633**, 906
Calchi Novati S 2010, *GRG* **42**, 2101
Calchi Novati S et al 2012, *in preparation*
Calchi Novati S and Mancini L 2011, *MNRAS* **416**, 1292
Calchi Novati S et al 2010, *ApJ* **717**, 987
Calchi Novati S et al 2009, *ApJ* **695**, 442
Calchi Novati S, de Luca F, Jetzer Ph, Mancini L and Scarpetta G 2008, *A&A* **480**, 723
Calchi Novati S et al 2007, *A&A* **469**, 115
Calchi Novati S, de Luca F, Jetzer Ph, and Scarpetta G 2006, *A&A* **459**, 407
Calchi Novati et al 2005, *A&A* **443**, 911
Crotts A 1992, *ApJ* **399**, L43
de Jong et al 2006, *A&A* **446**, 855
Dominik M 2010, *GRG* **42**, 2075
Dominik M et al 2010, *Astronomische Nachrichten* **331**, 671
Dominik M and Hirshfeld A 1996, *A&A* **313**, 841
Dominik M and Hirshfeld A 1994, *A&A* **289**, L31
Dong S 2007, *ApJ* **664**, 862
Gaudi S 2010, Refereed chapter in EXOPLANETS, edited by S. Seager, to be published as part of the Space Science Series of the University of Arizona Press (Tucson, AZ), (*Preprint* arXiv:1002.0332)
Gould A 2001, *PASP* **113**, 903
Gould A 2000, *ApJ* **535**, 928
Gould A 1996, *ApJ* **470**, 201
Gould A et al 2010, *ApJ* **720**, 1073
Ingrosso G, Calchi Novati S, de Paolis F, Jetzer Ph, Nucita A and Zakharov A 2011, *GRG* **43**, 1047
Ingrosso G, Calchi Novati S, de Paolis F, Jetzer Ph, Nucita A and Zakharov A 2009, *MNRAS* **399**, 219
Ingrosso G, Calchi Novati S, de Paolis F, Jetzer Ph, Nucita A, Scarpetta G and Strafella F, 2007, *A&A* **462**, 895
Jetzer Ph 1994, *A&A* **286**, 426
Lee C.-H. et al 2011, *accepted in AJ* (*Preprint* arXiv:1109.6320)
Mancini L, Calchi Novati S, Jetzer Ph and Scarpetta G 2004, *A&A* **427**, 61
Moniez M 2010, *GRG* **42**, 2047
Paczynski B 1986, *ApJ* **304**, 1
Paulin-Henriksson S et al 2003, *A&A* **405**, 15
Riffeser A, Seitz S and Bender R 2008, *ApJ* **684**, 1093
Riffeser A, Fliri J, Bender R, Seitz S and Gössl C 2003, *ApJ* **599**, L17
Sumi T 2011, in XV International Conference on Gravitational Microlensing: Conference Book, Bozza V, Calchi Novati S, Mancini L and Scarpetta G eds (*Preprint* arXiv:1102.0452)
Sumi T et al 2011, *Nature* **473**, 349
Sumi T et al 2010, *ApJ* **710**, 1641
Tisserand P 2007, *A&A* **469**, 387
Udalski A 2011, in XV International Conference on Gravitational Microlensing: Conference Book, Bozza V, Calchi Novati S, Mancini L and Scarpetta G eds (*Preprint* arXiv:1102.0452)
Wyrzykowski Ł et al 2011a, *MNRAS* **416**, 2949
Wyrzykowski Ł et al 2011b, *MNRAS* **413**, 493
Wyrzykowski Ł et al 2010, *MNRAS* **407**, 189
Wyrzykowski Ł et al 2009, *MNRAS* **397**, 1228
Zaritsky D and Lin D 1997, *AJ* **114**, 2545
Zhao H and Evans N 2000, *ApJ* **545**, L35
Zhao H, Ibata R, Lewis G and Irwin M 2003, *MNRAS* **339**, 701

Journal of Physics: Conference Series **354** (2012) 012002 doi:10.1088/1742-6596/354/1/012002

Overview of the LARES Mission: orbit, error analysis and technological aspects.

Ignazio Ciufolini[1], **Antonio Paolozzi**[2], **Claudio Paris**[2]

[1] Dip. Ingegneria dell'Innovazione, Università del Salento and INFN Sezione di Lecce, Italy
[2] Scuola di Ingegneria Aerospaziale, Sapienza Università di Roma, Italy

E-mail: `ignazio.ciufolini@unile.it`

Abstract. LARES (LAser RElativity Satellite), is an Italian Space Agency (ASI) mission to be launched beginning of 2012 with the new European launch vehicle, VEGA; the launch opportunity was provided by the European Space Agency (ESA). LARES is a laser ranged satellite; it will be launched into a nearly circular orbit, with an altitude of 1450 km and an inclination of 69.5 degrees. The goal of the mission is the measurement of the Lense-Thirring effect with an uncertainty of few percent; such a small uncertainty will be achieved using LARES data together with data from the LAGEOS I (NASA) and LAGEOS II (NASA and ASI) satellites, and because GRACE mission (NASA-CSR and DLR-GFZ) is improving Earth's gravity field models. This paper describes LARES experiment along with the principal error sources affecting the measurement. Furthermore, some engineering aspects of the mission, in particular the structure and materials of the satellite (designed in order to minimize the non-gravitational perturbations), are described.

1. Introduction

1.1. Dragging of Inertial Frames and the Lense-Thirring Effect

In Einstein's Theory of General Relativity, local inertial frames are influenced and dragged by the distribution and flow of mass energy in the Universe. In particular, they are dragged by the motion and rotation of nearby bodies; this phenomenon is known as Frame Dragging [1, 2].

Frame Dragging influences both the motion of bodies and the flow of time around spinning bodies. In any rigid frame non rotating relative to fixed stars, synchronization of two clocks orbiting around a spinning body is not possible if the clocks are moving in opposite directions on the same path; indeed the clock traveling on the orbit in the opposite direction of the rotation of the central spinning body will experience a time delay with respect to the clock traveling in the same direction of the rotation of the spinning body. Similarly, the orbital period of a particle co-rotating around a spinning body would be longer than the orbital period of a particle counter-rotating on the same orbit. [3, 4, 5, 6, 7].

In 1916 Willem de Sitter derived the Mercury perihelion precession due to the Sun angular momentum, on the basis of General Relativity. In 1918 Josef Lense and Hans Thirring gave a description of the frame dragging effect of a test particle orbiting around a spinning body; the so-called Lense-Thirring effect predicts the precession of the nodal line of the orbit, because the plane of the orbit is dragged by the rotation of the spinning body. The de Sitter effect is the precession of the pericenter (Geodetic Precession), but it is due simply to the presence of the central mass, that could be either spinning or not spinning, while the Lense-Thirring effect requires the presence of a spinning mass. Frame-dragging can be usefully described by a formal analogy of General Relativity, in a weak gravitational field and for slow motion, with electrodynamics [1, 8].

1.2. Measurement of the Lense-Thirring Effect with Laser Ranged Satellites

The idea of measuring the Lense-Thirring effect using the node of two laser ranged satellites dates back to 1984 [9, 10]; the proposal was to use the LAGEOS satellite (launched in 1976) and a second satellite to be launched on an orbit with all the orbital parameters identical to the ones of LAGEOS, except for the inclination that should have been supplementary (i.e. 70 degrees). The supplementary inclination was needed in order to cancel out all the secular effects on the nodes of the two satellites due to the deviations of the Earth's gravitational field from spherical symmetry and in particular due to the Earth's even zonal harmonics. The LAGEOS II satellite (a replica of LAGEOS I) was launched in 1992, but it was not possible to launch it on an orbit with the needed 70 degrees inclination (LAGEOS II orbit inclination is 52.64 degrees). However, to cancel the uncertainties due to the first N-1 even zonal harmonics, it is possible to use the nodes of N laser-ranged satellites [11, 1, 12, 13], as a possible alternative to the concept of the supplementary inclination satellites. An alternative possibility is to use N observables instead of N satellites; so the nodes of LAGEOS I and LAGEOS II, and the perigee of LAGEOS II, provide 3 observables that can be used to cancel the effect of the uncertainty due to the J2 and J4 harmonics in order to measure the Lense-Thirring effect. The above mentioned method was proposed in 1989 [11], before the launch of LAGEOS II, and described in detail in 1996 [13]. With this technique it was possible a first observation of the Lense-Thirring effect, applying the gravity field model EGM96 [14, 15]; however improved models of the Earth's gravitational field provided by GRACE mission [16, 17, 18, 19] lead to a measurement of the Lense-Thirring effect with an uncertainty of about 10% [12, 20, 21, 22]. To further improve the measurement, and to reach an uncertainty of the order of 1%, a new laser ranged satellite is needed.

2. LARES, a new laser-ranged satellite on a lower orbit than LAGEOS I and II

The proposal of a third laser ranged satellite, to be called LAGEOS III, dates back to 1989 [23, 24]. The possibility of launching such a third LAGEOS replica was discussed in a number of papers and PhD dissertations [23, 24, 25, 26]; however the mass (about 400 kg) and the high altitude of the orbit (about 6,000 km) required an expensive launch vehicle. For this reason a lighter satellite (about 100 kg), called LARES, was later designed [27], nevertheless the high altitude of LARES was still somehow expensive to achieve.
In 2007 the possibility to use the qualification flight of the new European launcher VEGA to orbit LARES was proposed [28]. However, this launch will put LARES on a lower orbit (1450 km) than the originally planned one. The possibility of using a lower orbit was supported by further studies and works the next year [29]. Even if VEGA could achieve a 6000 km orbit, the objective of the qualification flight did not allow to exceed an altitude of 1500 km. For this reason LARES was redesigned to have a more favourable mass-to-surface ratio (M/S) than LAGEOS; indeed LARES M/S ratio will be about 2.6 times the LAGEOS M/S ratio. The greater M/S ratio would help reducing the errors due to surface perturbations such as the thermal thrust and the atmospheric drag.

3. LARES Orbit

A suitable orbit for LARES would:
- minimize the effect of even zonal harmonics on the motion of the node;
- have small uncertainties due to the effects of other harmonics on the precession of the node;
- be perturbed by Earth's gravitomagnetic field, in order to measure the Lense-Thirring effect;

A polar orbit was also studied for LARES [30]: for such an orbit the effects of all even zonal

harmonics on the node will be theoretically absent. But at the same time it was demonstrated [26] that the uncertainty in the K_1 tide (tesseral, $m = 1$, tide) would make a polar orbit unsuitable for the Lense-Thirring measurement, since the corresponding uncertainty on the secular precession of the node would be too large. A better choice is a quasi polar orbit, with an inclination less than or equal to 86 degrees, or larger than or equal to 94 degrees (for an altitude of about 1500 km); this kind of orbit would have a periodical nodal precession, due to its departure from 90 degrees of inclination, and is also less demanding in terms of orbital injection deviation compared to a polar one. The analysis of LAGEOS I and LAGEOS II showed that the effect of K_1 tide can be fitted using a periodical signal exactly at the nodal frequency [14, 12].
By using the nodal rates of LARES, LAGEOS I and LAGEOS II as three observables, it is possible to eliminate the uncertainty in the first two even zonal harmonics, and the Lense-Thirring effect; the remaining uncertainties are due to the even zonal harmonics with degree strictly higher than 4. It turns out that some values of the inclination of LARES would minimize the error in the measurement of the Lense-Thirring effect since they would minimize the error due to the uncertainty in the largest (not cancelled using the combination of the three observables) even zonal harmonic C_{60}. The error as a function of the inclination of LARES orbit is shown in Figure 1, for an altitude of 1500 km (that is the altitude that can be reached with VEGA launch): the error would be minimized by an inclination of 70 degree or 110 degrees, but any inclination between 60 and 86 degrees and between 94 and 120 degrees could be used. The above results have been calculated assuming a circular orbit (e=0) and considering only the effects of the first 5 even zonal harmonics, C_{20}, C_{40}, C_{60}, C_{80} and C_{100} (by including higher degree even zonal harmonics, the results of Figure 1 would only change slightly). The uncertainties in the spherical harmonics are considered equal to those of the EIGEN-GRACE02S Earth's gravity model [16]; EIGEN-GRACE02S was a preliminary 2004 model and by the time of the launch of LARES and of its data analysis, much improved Earth's gravity field models based on much longer data set of GRACE observations would be available.

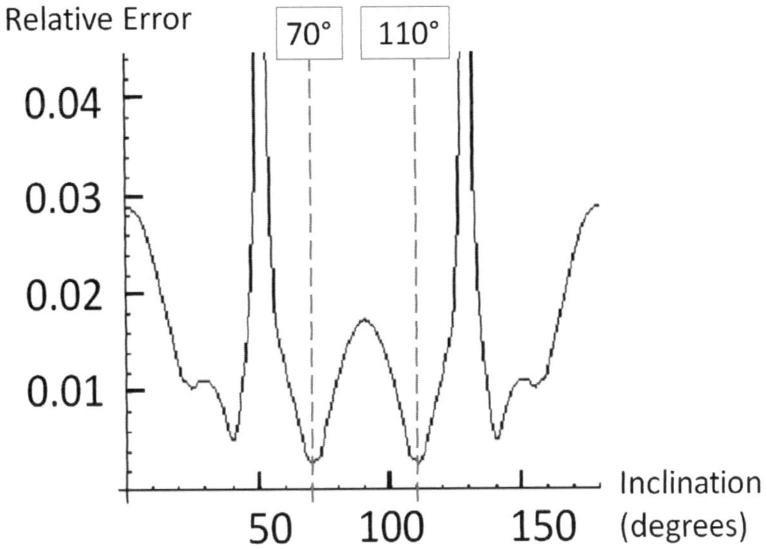

Figure 1. Relative error due to the largest even zonal harmonics, as function of the inclination of LARES orbit (for an altitude of 1500 km and zero eccentricity).

Other relevant orbital perturbations that affect the measurement are: tidal effect, particle drag and thermal drag.
The combination of three nodes makes possible to cancel the effect of medium and long period zonal tides ($l = 2$ and $m = 0$) together with the static C_{20} uncertainty and the uncertainty in the time-dependent secular variations $\dot{C}_{20}$, $\dot{C}_{40}$. As explained above, the effect of K_1 tide will be fitted for over a period equal to the LARES nodal period, and the introduced uncertainty will be small [23, 24, 26].
In regard to the thermal drag, the effect of the unmodeled perturbation would be reduced with respect to the LAGEOS satellite by the greater M/S ratio (as stated above, it would be about 2.6 times than LAGEOS M/S); furthermore, measurements of the thermal properties of the LARES materials are planned, to help modeling the thermal drag effect.
The higher M/S ratio would also help reducing the effect of particle drag; nevertheless, the effect of drag on the node of an orbit with nearly zero eccentricity and an altitude of 1450 km would be negligible (indeed, the effect is proportional to eccentricity, and will be zero for e=0)[11].

4. Analysis of Errors due to Gravitational Uncertainties

The largest gravitational perturbations on the orbit of the LAGEOS satellites and LARES are due to the non-sphericity of the Earth gravity field, described by the expansion of the potential in spherical harmonics [31, 11]. The spherical harmonics of even degree and zero order, e.g., the C_{20} harmonic describing the well known quadrupole moment of Earth, are the source of the only secular effects on the orbit.
As stated above, the error due to the uncertainties of the first two even zonal harmonics will be canceled using the nodes of three satellites; the error due to each even zonal harmonic of degree higher than 4 is considerably less than 1% and in particular that the error is substantially negligible for the even zonal harmonics of degree higher than 26. The total error in the measurement of the Lense-Thirring with LARES due to the even zonal harmonics has been calculated for the gravity field models EIGEN-GRACE02S (GFZ Potsdam, 2004) [16] and GGM02S (CSR 2004) [17], obtaining a value respectively of 1.4% and 2.1% [32]. The above mentioned models have been obtained in 2004 with data over less than 365 days of GRACE mission, therefore improved models would be available at the time of the LARES data analysis, based on longer GRACE observational periods. Moreover GOCE mission [33] is going to further improve the gravity field models, even if mainly the higher degree harmonics, whose contribution to the total error is only marginal. Indeed the contribution of high degree spherical harmonics of the gravity field to the satellite nodal precession decrease as the inverse power of the semimajor axis to the degree of the even zonal harmonic, thus quickly decreasing with the degree [31, 22].

5. Technological Aspects of LARES Satellite

LARES will be launched with the VEGA maiden flight; the satellite will be put on a nearly circular orbit, with a semimajor axis of 7830 km and an inclination of 69.5 degrees. Being on a much lower orbit than LAGEOS I and II, it is important that the design will minimize all the non gravitational perturbations, especially particle drag and thermal thrust (induced by the anisotropic thermal radiations from the satellite due to the anisotropic temperature distribution over the satellite surface). In Figure 2 is reported a photograph of the qualification model of LARES satellite. In the following subsection the principal design aspects are discussed.

5.1. Mass-to-Surface Ratio, thermal thrust and materials

The mass-to-surface (actually mass-to-cross-sectional-area) ratio, M/S, should be maximized in order to minimize the non gravitational perturbations. For a spherical satellite, such as LARES, a simple formula shows that M/S depends only on satellite radius, r, and satellite mean density,ρ:

$$\frac{M}{S} = \frac{4\rho r}{3} \tag{1}$$

However, the launch vehicle imposes some limits on the maximum mass of the satellite. For LARES, the mass limit was 400 kg, so the most obvious decision was to maximize the density of the satellite. The satellite body will be machined from one single sphere of Tungsten alloy THA-18N, composition 95% W, 5% Cu and Ni, nominal density ρ=18000 kg/m^3 [34, 35]. Higher density Tungsten alloys are available, that are magnetic alloys containing Fe; a non magnetic alloy (even if with a lower density) has been chosen to avoid interaction with the Earth magnetic field. Tungsten alloys have been preferred over other materials with even higher density, such as platinum, osmium and iridium, for both economical reasons (platinum price is about the same as gold, iridium is priced 0.75 times than gold), and technological reasons (availability, workability and procurement time).
Another solution to increase the M/S ratio is the adoption of conical cavities to house the CCRs; indeed, while the CCR design is inspired to the LAGEOS satellites, the cylindrical cavities used on LAGEOS satellite would decrease the M/S ratio of LARES.

To minimize perturbations due to the thermal thrust, the satellite is made of a single piece of metal. This way thermal contact conductances are eliminated on the satellite body thus reducing drastically temperature differences. To reduce risk of differential charging all the metal components, i.e. screws and CCR retaining rings, are made of the same alloy [35, 36, 37].
Given the aforementioned decision about the materials and the design, LARES satellite will have a radius of about 18 cm and a mass of about 390 kg; indeed it will become the densest known single object orbiting in the Solar System.

5.2. Cube Corner Reflectors
The most critical components of the LARES satellite are the Cube Corner Reflectors (CCRs). The CCRs are designed to redirect the laser pulses back to the ground station, to allow Laser Ranging. The dihedral angles between back faces of a CCR is exactly 90 only if the target motion is zero or negligible (this is the case of the CCR used for Lunar Laser Ranging); in the case of a satellite in Earth orbit, the dihedral angle should have a small offset (Dihedral Angle Offset, DAO) to spread the reflected signal over a wider area to compensate for the satellite motion. LARES CCRs have a DAO of 1.5 arcsec, with a tolerance of +/- 0.5 arcsec.
LARES carries 92 CCRs. A higher number of CCRs would have reduced slightly the M/S ratio, and increased the thermal thrust perturbation; moreover, even if a higher number of CCRs would have increased the reflective area, a higher number of CCRs reflecting at the same time the laser pulse would have decreased the laser ranging accuracy. A spare unit of a CCR made for LARES mission is shown in Figure 3.

5.3. CCRs Mounting system
On the LAGEOS satellites, the CCRs and their retainer rings (made of aluminium) were found to be the major cause of unmodeled thermal thrust. LARES will employ a CCR mounting system inspired to the LAGEOS system (that has been proven to be reliable), with the main difference that the metal mounting rings and the screws are made of the same tungsten alloy as the satellite body. Methods to improve thermal conductivities between the CCRs and the satellite body are possible but were not recommended. For instance the application of a coating with a high infrared emissivity on the CCR cavity would have increased the thermal flux towards the CCR back faces reducing the temperature difference between CCR and satellite body. On the other hand this flux would have possibly increased the axial thermal gradients on the CCR

Figure 2. Qualification Model (QM) of LARES, with four CCRs installed. LARES QM is made of the same material and has the same dimensions of LARES Flight Unit. Between the four CCRs there is clearly visible one of the four hemispherical cavities that interface the Separation System.

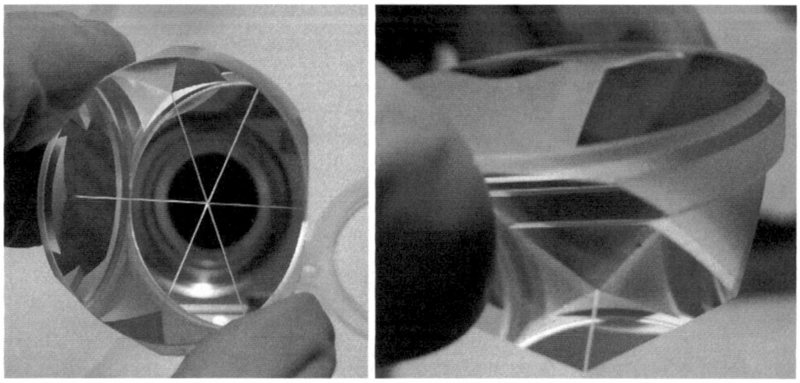

Figure 3. Front view (left) and side view (right) of a LARES CCR (spare unit).

causing a dangerous CCR deformation that would have changed its Far Field Diffraction Pattern ultimately reducing the energy at the ground station of the returning laser pulse.

5.4. Separation System

To avoid unmodeled effects due to particle drag, LARES satellite has been designed avoiding any component protruding from the spherical surface. This requirement called for a dedicated separation system. The system holds the satellite by 4 brackets with spherical heads interfacing 4 corresponding spherical slots spaced 90 degrees along the equatorial plane of the satellite. Each bracket is operated by a non-pyrotechnical actuator; the four mechanisms will be operated simultaneously to release the satellite once in orbit. A spring will then push LARES away from the VEGA upper stage, with a nominal separation speed of 0.75 m/s. The system has been designed to be robust against failures: in case of failure of one mechanism, the remaining three will allow to release the satellite, even if in this case the polar spin axis will be lost.

Figure 4. A 3D model of LARES Separation System, showing the four brackets holding the satellite.

6. Conclusions

The particular design of LARES satellite, together with the improved gravity field models from GRACE mission, will improve the accuracy in the measurement of Earth gravitomagnetic field. The VEGA launch vehicle imposes some constraints on LARES mass and orbit; however the orbital parameters, in particular eccentricity and inclination, are chosen to reduce the errors due to the effect of even zonal harmonics and Earth tides. The robust mission design allow big orbital injection errors if the final orbit has an altitude higher than 1200 km. The semimajor axis has been decided after a trade off with the launch vehicle responsible.

References

[1] Ciufolini I and Wheeler J A 1995 *Gravitation and Inertia* (Princeton: Princeton University Press)
[2] Einstein A 1913 *Letter to Ernst Mach.* (Zurich, 25 June) in Misner CW, Thorne K S, and Wheeler J A, 1973 *Gravitation* (San Francisco: Freeman)
[3] Zeldovich Ya B and Novikov I D 1971 *Relativistic Astrophysics*, **I**, Stars and Relativity (Chicago: Univ. Chicago Press)
[4] Landau L D and Lifshitz E M 1971 *The Classical Theory of Fields* 3rd rev. English edn. (London: Pergamon)
[5] Ciufolini I and Ricci F 2002 Time delay due to spin and gravitational lensing. *Classical and Quantum Gravity* **19**
[6] Ciufolini I and Ricci F 2002 Time delay due to spin inside a rotating shell. *Classical and Quantum Gravity* **19**
[7] Ciufolini I, Ricci F, Kopekin S and Mashhoon B 2003 On the gravitomagnetic time delay *Phys. Lett. A* **308** 101/109
[8] Thorne K S, Price R H and Macdonald D A 1986 *The Membrane Paradigm* (NewHaven: Yale Univ. Press)
[9] Ciufolini I 1984 Theory and Experiments in General Relativity and other Metric Theories *PhD Dissertation* (Univ. of Texas)
[10] Ciufolini I 1986 *Phys. Rev. Lett.* **56** 278
[11] Ciufolini I 1989 *Int. J. Mod. Phys. A* **4** 3083
[12] Ciufolini I and Pavlis E 2004 *Letters to Nature* **431** 958
[13] Ciufolini I 1996 *Nuovo Cimento A* **109** 1709
[14] Ciufolini I, Pavlis E C, Chieppa F, Fernandes-Vieira E and Perez-Mercader J 1998 *Science* **279** 2100
[15] Ciufolini I, Chieppa F, Lucchesi D and Vespe F 1997 *Class. and Quantum Grav.* **14** 2701
[16] Reigber C, Schmidt R, Flechtner F, Koenig R, Meyer U, Neumayer K H, Schwintze P, and Zhu S Y 2005 An Earth gravity field model complete to degree and order 150 from GRACE: EIGEN-GRACE02S *J. Geodyn.* **39** 1
[17] Reigber Ch, Flechtner F, Koenig R, Meyer U, Neumayer K, Schmidt R, Schwintzer P, and Zhu S 2002 GRACE Orbit and Gravity Field Recovery at GFZ Potsdam - First Experiences and Perspectives *Eos. Trans. AGU* 83(47) Abstract G12B-03 (Fall Meet. Suppl.)
[18] Tapley B D 2002 The GRACE Mission: Status and Performance Assessment *Eos. Trans. AGU* 83(47) Abstract G12B-01 (Fall Meet. Suppl.)
[19] Watkins M M, Yuan D, Bertiger W, Kruizinga G, Romans L, and Wu S 2002 GRACE Gravity Field Results from JPL *Eos. Trans. AGU* 83(47) Abstract G12B-02 (Fall Meet. Suppl.)
[20] Ciufolini I, Pavlis E and Peron R 2006 *New Astronomy* **11** 527
[21] Ries J C, Eanes R J and Watkins M M 2008 Confirming the Frame-Dragging Effect with Satellite Laser Ranging *Proc. of 16th International Workshop on Laser Ranging* (Poznan: 13-17 October)
[22] Ciufolini I, Pavlis E C, Ries J C, Koenig R, Sindoni G, Paolozzi A, Eanes R and Newmayer H 2009 Gravitomagnetism and its Measurement with Laser Ranging to the LAGEOS satellites and GRACE Earth Gravity Models, in *John Archibald Wheleer and General Relativity* (Springer)
[23] Tapley B, Ries J C, Eanes R J, and Watkins M M 1989 *NASA-ASI Study on LAGEOS III* (Austin: CSR-UT publication n CSR-89-3)
[24] Ciufolini I et al 1989 *ASI-NASA Study on LAGEOS III* (Rome: CNR)
[25] Ries J C Simulation of an experiment to measure the Lense-Thirring precession using a second LAGEOS satellite 1989 *PhD Dissertation* (Univ. of Texas, Austin)
[26] Peterson G E 1997 Estimation of the Lense-Thirring precession using laser-ranged satellites *PhD Dissertation* (Univ. of Texas, Austin)
[27] Ciufolini I, Paolozzi A, Pavlis E et al 1998 *LARES phase A study for ASI* (ASI Tech. Rep.)
[28] Paolozzi A 2007 Private communication
[29] Ries J 2007 Private communication
[30] Lucchesi D M, Paolozzi A 2001 A Cost Effective Approach for LARES Satellite *Proc. of XVI AIDAA Conference* pp 1-14 Paper n 111 (Palermo: September)
[31] Kaula W M 1966 *Theory of Satellite Geodesy* (Waltham: Blaisdell)
[32] Ciufolini I and Matzner R A (eds) 2010 *General Relativity and John Archibald Wheeler* Astrophysics and Space Science Library **367** (Springer Science + Business Media B.V.)
[33] Rummel R 2003 GOCE - its status and promise *Geophysical Research Abstracts* **5** (CD) Abstract EAE03-A-09628
[34] Ciufolini I, Paolozzi A, Dell'Agnello S, Peroni I et al 2007 The design of LARES, a Satellite for Testing General Relativity *Proc. of 58th International Astronautical Congress* IAC-07-B4.2.07 (Hyderabad: 24-28 September)
[35] Paolozzi A, Ciufolini I, Felli F, Brotzu A, Pilone D 2009 Issues on LARES Satellite Material *Proc. of 60th International Astronautical Congress* IAC-09-C2.4.5 (Daejeon: 12-16 October)

[36] Paolozzi A, Ciufolini I, Vendittozzi C 2011 Engineering and scientific aspects of LARES satellite *Acta Astronautica* **69** pp 127-134 (Elsevier)

[37] Paolozzi A, Ciufolini I, Vendittozzi C, Passeggio F, Caputo L, Caputo G 2009 Technological Challenges for Manufacturing Lares Satellite *Proc. of 60th International Astronautical Congress* IAC-09.C2.I.5 (Daejeon: 12-16 October)

Journal of Physics: Conference Series **354** (2012) 012003 doi:10.1088/1742-6596/354/1/012003

High-energy observations of black hole binaries with the INTEGRAL satellite

Melania Del Santo

INAF/IASF-Roma, via del Fosso del Cavaliere 100, 00133, Roma, Italy

E-mail: melania.delsanto@iasf-roma.inaf.it

Abstract. Black-hole binaries are important sources through which studying accretion onto compact objects. In the X/γ-ray domain, these objects show several and complex spectral behaviours and transitions. Based on *INTEGRAL* observations collected during the last eight years, we have now a new view on the high energy emission of black-hole binary. An additional component above 200 keV has been observed in a few systems, during either hard/intermediate or low/hard states. The nature of this hard-tail is still debated, as also the one observed in soft states. However, among a number of models, it is usually attributed to the presence of a small fraction of non-thermal electrons in a hot-Comptonising plasma. I review the high energy emission from black hole binary systems and report on some *INTEGRAL* observations of three different objects: 1E 1740.7–2942, GX 339–4, Cyg X-1.

1. Introduction

When a compact object, like black hole (BH) or neutron star (NS), accretes matter from a normal star, the inner part of the accretion flow (disc) becomes a bright X-ray emitter. In such a system, i.e. X-ray binary, matter is transferred through the inner Lagrangian point, when the donor is less massive than the compact object (Low Mass X-ray Binary, LMXB) or the compact object captures mass from the wind of a massive star, usually B, O, Be or supergiant (High Mass X-ray Binary, HMXB). In both cases, the fate of the accreting matter depends on its angular momentum and physical processes by which it loses this angular momentum, and on the radiation processes involved (see [1] for a review on the accretion power in astrophysics).

In the early 1970s and for the next two decades, most of the modeling effort of accretion flows onto neutron stars and black holes was based on two assumptions: i) the accretion flows were assumed to be loosing angular momentum at high rates because of an unspecified process, with the effective kinematic viscosity typically taken to be proportional to the pressure; ii) the radiation processes were assumed to be very efficient, so that the resulting accretion flows were relatively cool, in the form of geometrically thin accretion disks (see [2]).

During the last decade, theoretical models of accretion flows onto compact objects became increasingly more sophisticated. The two major developments were: the identification of a magneto-hydrodynamic instability in differentially rotating flows [3, 4] and the discovery of new stable solutions to the hydrodynamic equations that describe radiatively inefficient accretion flows [5]. In these solutions, the electrons and ions have different and high temperatures, the accretion flows are geometrically thick, and most of their potential energy is not radiated away but is rather advected towards the compact objects (Advection-Dominated Accretion Flows, ADAFs). See [6] for a review on accreting neutron stars and black holes.

An important ingredient in the models of accretion flows onto compact objects is the interaction of the flows with the objects themselves. This is the region in the accretion flows where most of the X-ray and γ-ray radiation is produced and observed with high-energy telescopes. Clearly, the interaction depends on whether the collapsed object is a black hole or a neutron star. Observational features such as pulsations and/or type-I bursts are proofs of the existence of a surface: in these cases the nature of the compact object is definitely unveiled as being a neutron star.

1.1. Spectral States of Black Hole Binaries

Black hole binaries (BHB) are known to show different spectral states in the X and γ-ray domain. The two main spectral states are the Low/Hard State (LHS), with the high energy spectrum described by a cut-off power-law (typically $\Gamma \sim$1.5 and $E_{cut} \sim$100 keV), and the High Soft State (HSS) with a thermal component peaking at few keV and the high energy power-law much softer ($\Gamma > 2.2$) [7]. Usually, this spectral variability is interpreted as due to changes in the geometry of the central parts of the accretion flow (see [8] for a recent review). In the LHS the standard geometrically thin and optically thick disc [2] would be truncated far away from the last stable orbit. In the innermost parts, a hot accretion flow is responsible for the high energy emission, via thermal Comptonisation of the soft photons coming from the truncated disc. In the HSS, the optically thick disc would extend close to the minimum stable orbit producing the dominant thermal component. The weak non-thermal emission at higher energy is believed to be due to up-scattering of the soft thermal disc emission in active coronal regions above the disc [9]. A number of intermediate states with spectral parameters in between the two canonical states are also observed. A brief review on the radiative processes responsible for X-ray emission in hard (low) and soft (high) spectral states of black-hole binaries is [9]. See [11] as more recent and exhaustive review on BH binaries.

1.2. Black hole candidates

In the X and γ-ray domain, BHBs manifest themselves in a number of different spectral/temporal states [11]. Up to now, 18 object are dynamically confirmed black hole binaries [11], while more than 30 objects are black hole candidates (BHC). Certain characteristic X–ray properties of these established BHBs are often used to identify a BHC when the radial velocities of the secondary cannot be measured (e.g., because the secondary is too faint). However, these signatures have been observed in a number of systems known to contain a NS primary[10] and definitely cannot be used for the BH identification. This is not surprising since the X–ray spectral/temporal properties originate in an accretion flow that is expected to be fairly similar whether the primary is a BH or a weakly magnetized NS.

Of special importance is the mass function, $f(M) \equiv P_{\rm orb}K_2^3/2\pi G = M_1 \sin^3 i/(1+q)^2$. The observables on the left side of the equation are the orbital period, $P_{\rm orb}$, and the half–amplitude of the velocity curve of the secondary, K_2. On the right, the quantity of most interest is the BH mass, M_1; the other parameters are the orbital inclination angle, i, and the mass ratio, $q \equiv M_2/M_1$, where M_2 is the mass of the secondary. Thus a secure value of the mass function can be determined for a quiescent X–ray nova or an HMXB by simply measuring the radial velocity curve of the secondary star. An inspection of the equation for $f(M)$ shows that the value of the mass function is the absolute minimum mass of the compact primary. Thus, for 12 of the 18 BHBs, the very secure value of f(M) alone is sufficient to show that the mass of the compact X–ray source is at least 3 $M_\odot$, which is widely agreed to exceed the maximum stable mass of a neutron star (NS) in GR [12, 13].

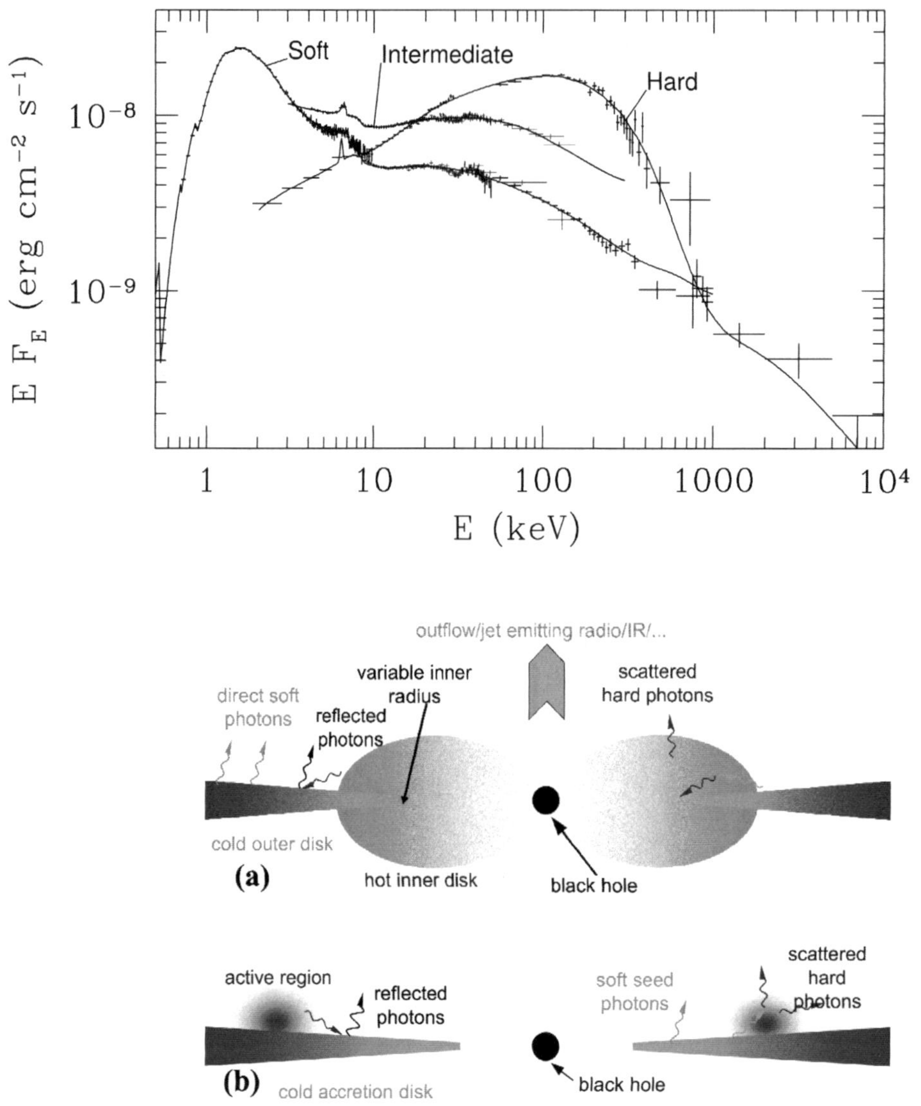

Figure 1. *Top*: Energy spectra in three states of Cyg X-1[7]. *Bottom*: (a) A schematic representation of the likely geometry in the hard state, consisting of a hot inner accretion flow surrounded by optically-thick accretion disk. The hot flow constitutes the base of the jet (with the counter-jet omitted from the figure for clarity). The disk is truncated far away from the minimum stable orbit, but it overlaps with the hot flow. The soft photons emitted by the disk are Compton up-scattered in the hot flow, and emission from the hot flow is partly Compton-reflected from the disk. (b) The likely geometry in the soft state consisting of flares/active regions (non-thermal electrons) above an optically-thick accretion disk extending close to the minimum stable orbit. The soft photons emitted by the disk are Compton up-scattered in the flares, and emission from the flares is partly Compton-reflected from the disk [9]. (Credit: Zdziarski 2000 and Zdziarski & Gierliński 2004).

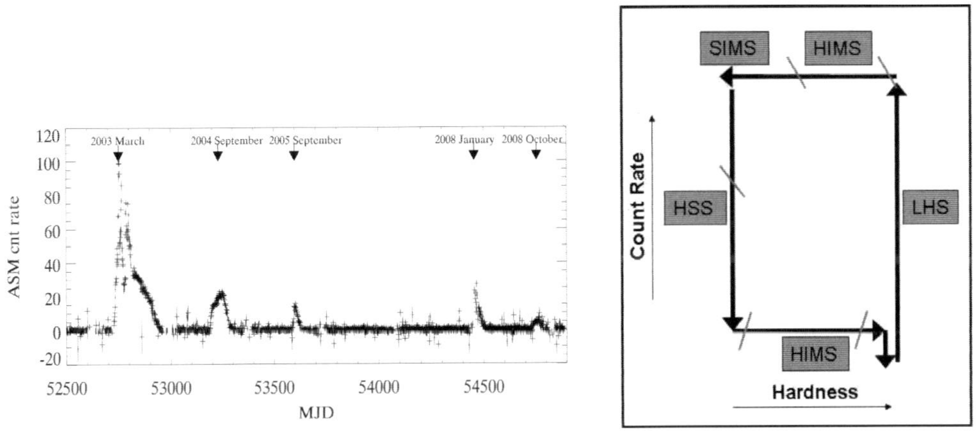

Figure 2. *Left*: *RXTE*/ASM 1.5-12 keV light curves of H 1743–322 from 2003 until the end of 2008. Five outbursts at different flux levels have been observed by *RXTE* until 2008. *Right*: Sketch of the "q-track pattern" indicating the evolution through all the different spectral states of a typical outburst of transient BHCs.

1.3. Evolution of a transient LMXB with black hole as compact object

During the outbursts of transient black hole candidates (BHCs), in addition to the large changes in X-ray luminosity (Fig. 2), marked variations are observed in the properties of the timing and the energy spectrum often on very short time scales (see e.g. [15]). A BHC spends its time mostly in a quiescent state at low flux level ($< 10^{32}$–10^{33} erg s^{-1}, e.g. [16]). When the outburst begins, the luminosity of the source increases and the X-ray spectrum is typical of the LH state (see previous section). The radio emission in this state indicates the presence of steady jets, while the power spectrum is dominated by a strong band limited noise (>30% fractional rms). Then the outburst evolves as the source increases its luminosity and its spectrum starts to change: the soft thermal component appears and becomes more important, the energy peak of the emission softens and the photon index of the hard component steepens ($\sim$ 2.0-2.5). Two different states with these spectral characteristics have been defined: the hard intermediate state (HIMS) and the soft intermediate state (SIMS). The characteristics of these two states are quite complex: the changes can be established mostly by the timing properties [17] and also by the ejection of relativistic jets associated to the transition from HIMS to SIMS [18]. After the SIMS, the source enters a state where the X-ray spectrum is dominated by the emission of the soft thermal component (the so called HS state). A non-thermal power law tail is also present without any detectable cutoff, while the power spectrum is characterised by a low-level (1-2% fractional rms) variability. Then the flux starts to decrease, most likely following a parallel decrease in accretion rate. At some point, a reverse transition is started and the path is followed backwards all the way to the LHS and then to quiescence. As mentioned above, the luminosity level of this back-transition is always lower than that of the corresponding forward-transition. The hardness-intensity diagram of transient BHCs shows such a basic pattern, the so-called "*q-track*" pattern ([18, 17]; see Fig. 2, *right*).

2. The *INTEGRAL* satellite

The ESA *INTEGRAL* observatory (International Gamma-Ray Astrophysics Laboratory, [19]) was selected in June 1993 as medium-size scientific mission within ESA Horizon 2000 programme. *INTEGRAL* is dedicated to the fine spectroscopy (2.5 keV FWHM @ 1 MeV) and imaging

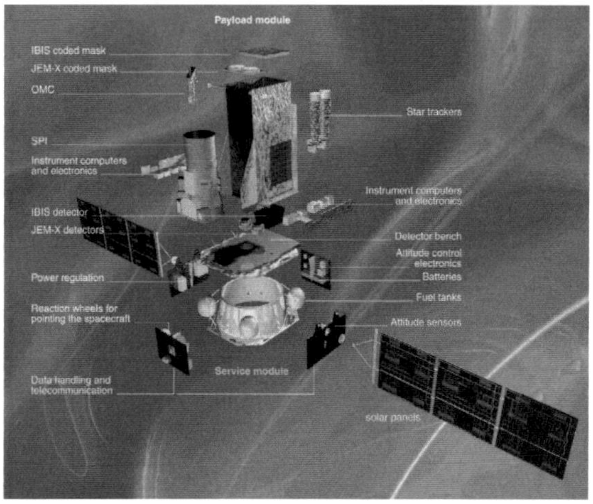

Figure 3. The INTEGRAL Observatory (Courtesy of ESA).

(angular resolution: 12' FWHM) of gamma-ray sources in the energy range 15 keV–10 MeV. While the imaging is carried out by the imager IBIS[20], the fine spectroscopy if performed by the spectrometer SPI [21]. Two coaxial monitors are dedicated to the X-ray (JEM-X, 3-35 keV[22]) and optical (OMC[23]) counterparts observation

SPI, IBIS and JEM-X are coded mask telescopes. This technique is used when photons focusing become impossible by using standard grazing technique (see Skinner & Connell 2004[24] for details). *INTEGRAL* was launched from Baikonur on 2002, October 17th and inserted into an highly eccentric orbit lasting 3 days. The data are received by the *INTEGRAL* Mission Operation Centre (MOC) in Darmstadt (Germany) and relayed to the INTEGRAL Science Data Center (ISDC[25]) which provide the final consolidated data products to the observes and later archived for public use. The proprietary data become public one year after distribution to single observation PIs.

The IBIS (Imager on Board *INTEGRAL* Spacecraft) imaging instrument is optimized for survey work with a large ($29° \times 29°$) field of view (FOV) with excellent imaging and spectroscopy capability. The data are collected with the low-energy detector, ISGRI (INTEGRAL Soft Gamma-Ray Imager[26]), working in the energy range 17-500 keV. A sequence of IBIS survey catalogs has been published at regular intervals as more data have become available [27, 28, 29, 30]. In the last catalogue (on 2010 [30]), 700 high-energy sources have been provided. Among them 22 are identified as black hole binaries or candidates. Up to now, more than 25 BHB/BHC have been observed by *INTEGRAL*: it is interesting to note that above 150 keV (up to 300 keV) the point sources population is totally dominated by these sources.

3. The microquasar 1E1740.7–2942

1E 1740.7–2942 is a bright hard X-ray source located at less than one degree off the Galactic Centre [31], classified as Black Hole Candidate [32] . When in 1992 [33] it was discovered a double sided radio jet reaching large angular distances from the core ($\sim$ 1'), the "microquasar" class was born with 1E 1740.7–2942 as its first member.

All observations performed so far revealed that 1E 1740.7–2942 spends most of the time in the canonical LH state of BHC. In a few occasions, soft spectral states have been observed during 1E 1740.7–2942 low flux levels [34]. Simultaneous *INTEGRAL* and *RXTE* broad-band spectral study performed in 2003 (Fig. 4, *left*) reported on an intermediate/soft spectral state (Fig. 4, *right*) occurred just before the source quenching [35]. One interpretation is a simple model

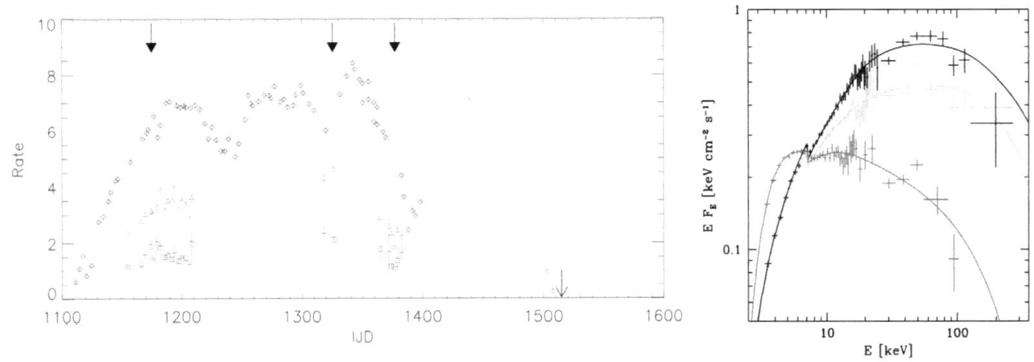

Figure 4. *Left*: Lightcurves from PCA (8–25 keV, blue diamonds), and from IBIS/ISGRI in the two bands 30–50 keV (orange triangles) and 50–80 keV (red squares). Solid arrows indicate the three periods used for the spectral analysis. Simple arrow indicates when the source started to be no longer visible by IBIS (beginning 2004). *Right*: The observed spectra, as fitted by a thermal Comptonization model [35]. The green and black spectra are canonical low/hard states; the red one is a soft/intermediate state.

with two simultaneous, independent accretion flows: a thin disk and a hot halo. A drop in the accretion rate affecting both flows would propagate through the halo immediately but might take up to several weeks to propagate through the disk. While the inner halo is thus temporarily depleted compared to the disk, a temporary soft state is expected [34].

3.1. High-energy spectral study

In Fig. 5 (*left*), I show the temporal behaviour of 1E 1740.7–2942 as observed with IBIS/ISGRI in 2003, 2004, 2005. Excluding occasional soft states, two main states have been observed during these three years: the canonical LH state with a mean flux of $\sim$ 50 mCrab and a "dim" state during which 1E 1740.7–2942 was at 2 mCrab flux level [36].

We averaged all IBIS/ISGRI and SPI Low/Hard spectra (from 20 keV) collected during the whole analysed period [37]. The averaged spectrum (roughly 2 Ms of exposure time) allowed us to achieve a better statistics at high-energy and eventually give better constraints to the spectral parameters. For the first time the continuum of the LH state of 1E 1740.7–2942 has been measured up to $\sim$600 keV. When adjusted with a single Comptonisation model (COMPTT [38]), an additional component is strongly required to fit the data above 200 keV. This high energy component has been observed in several BHCs [39, 40], usually during high soft spectral states, and explained as Compton up-scattering by a non-thermal electrons population [9]. Alternatively, we show that a model composed by two thermal Comptonisations provides an equally representative description of the data: the temperature of the first population of electrons (kT_1) results as $\sim$30 keV while the second, kT_2 is fixed at 100 keV (Fig. 5, *right*). This two temperature model could either correspond to two distinct heating mecanisms/regions or reflect the presence of a gradient of temperature in the Comptonising plasma [41].

3.2. The annihilation line

In 1990, the SIGMA telescope on-board *Granat* detected a broad line around the electron-positron annihilation energy [42, 43]. This transient feature appeared clearly during a 13 hours observation and then possibly in two further occasions but at a less significant level. Numerous works dedicated to similar line searches have followed and all led to negative conclusions (see

Journal of Physics: Conference Series **354** (2012) 012003 doi:10.1088/1742-6596/354/1/012003

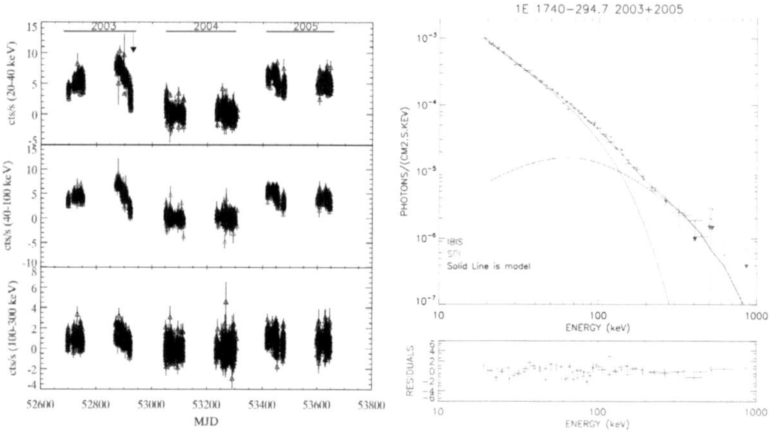

Figure 5. *Right*: 1E 1740.7–2942 temporal behaviour observed with IBIS/ISGRI in 2003, 2004, 2005 in 3 different energy bands. *Left*: 1E 1740.7–2942 spectrum observed by IBIS and SPI for 2003 + 2005 observations. Solid line is a two comptonisation model, with hotter temperature fixed to 100 keV. The optical depths are typical τ_1=1.6 and τ_2=2.2, respectively.

for example [44] and references therein). In this context, it has been interesting to perform a deep analysis of SPI data in this energy domain, and to seek for any feature around 511 keV associated with 1E 1740.7–2942.
The superior energy resolution of the SPI telescope allows for a specific dedicated study of this topic. Indeed, for the first time, an instrument is capable to look for a narrow feature in this particular source. During the first year of observations, no evidence for point source emission at 511 keV has been detected with SPI. The upper limit at 3.5 σ level is 1.6×10^{-4} ph cm^{-2} s^{-1} for a narrow line [45] while the IBIS data set a 2 σ upper limit of 4.1×10^{-4} ph cm^{-2} s^{-1} in the 535–585 keV energy band for an exposure time equal to 1.5 Ms [46].

Even though the complexity of the considered region makes it difficult to attribute firmly the detected emission to 1E 1740.7–2942, the presence of photons with energy greater than several hundreds of keV in a more or less persistent way (something as half or 2/3 of the time), together with previously reported annihilation emission, support a scenario in which 1E 1740.7–2942 is a source of positrons. Indeed, as proposed by van Oss & Belyanin (1995), a plasma detected with a temperature much lower than 1 MeV is able to produce positrons through photon-photon absorption. The basic argument is that the hard X-ray emission comes from the regions close to the central black hole, where the gravity field is very strong. The high local temperature is thus lowered, leading to an observed value far from the relativistic domain, while pairs are created in the innermost disk and driven away. Annihilation outbursts could occur when the accretion flow intercepts the pair wind [47]. Concerning the 511 keV line itself, no feature, broad or narrow, transient or persistent, has been found by *INTEGRAL* from point sources, confirming the rare occurrence of such a phenomenon in line with numerous different studies already performed on this topics.

4. The transient black hole binary GX 339-4

Since its discovery [48], the X-ray binary GX 339−4 has been thoroughly studied at all wavelength from radio to gamma-rays. Its star companion is still unknown, even though upper limit on the optical luminosity allowed to classify the source as Low Mass X-ray Binary (LMXB; [49]). Classified as BHC [50], GX 339−4 is a transient source spending long periods in outburst.

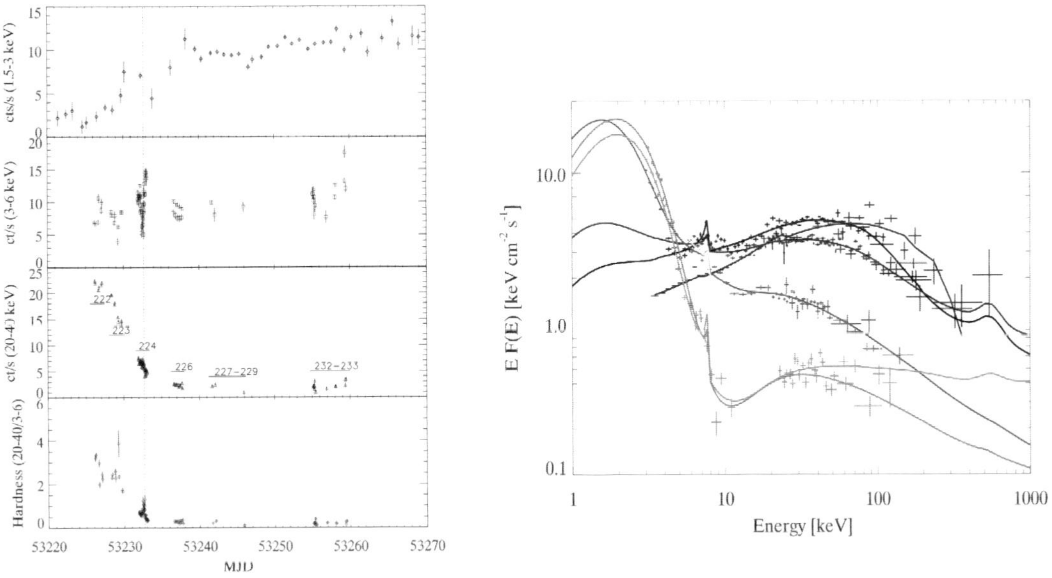

Figure 6. *Left*: GX 339−4 count rate in the energy ranges 1.5-3 keV, 3-6 keV, 20-40 keV with *RXTE*/ASM, JEM-X and IBIS/ISGRI, respectively. The related *INTEGRAL* orbits are also marked on the IBIS light curve. IBIS/ISGRI to JEM-X hardness ratio is shown in the bottom panel. *Right*: Joint JEM-X, IBIS and SPI unfolded energy spectra (Model: EQPAIR) of GX 339−4 averaged during different spectral states in 2004: HIMS for spectra 1 (black), 2 (blue), 3 (red); SIMS for 4 (yellow); HSS for spectra 5 (green) and 7 (orange). In order to give a comparison with a pure Low/Hard state spectrum, we show a spectrum (violet) reported in [56].

The long-term variability of GX 339−4 (1987-2004) is extensively presented in [51]. *INTEGRAL* has observed part of the 2004 outburst (August-September; Fig. 6, *left*) and reported on different spectral transitions [52].

4.1. The 2004 outburst as observed by INTEGRAL

In the period 2004 August 9th–September 11th *INTEGRAL* observed the GX 339−4 outburst (Fig. 6, *left*) during different spectral states, namely the hard/intermediate, soft/intermediate and the high/soft states. Joint JEM-X, IBIS and SPI collected during the different states of GX 339−4 have been fitted EQPAIR [53] (Fig. 6, *right* and Tab. 1). In the EQPAIR model, emission of the disc/corona system is modeled by a spherical hot plasma cloud with continuous acceleration of electrons illuminated by soft photons emitted by the accretion disc. At high energies the distribution of electrons is non-thermal, while at low energies a Maxwellian distribution with temperature kT_e is established. Our detailed spectral analysis of time averaged spectra at different stages of the transition confirms that the spectral transition in GX 339−4 is driven by changes in the soft cooling photon flux in the corona associated with an increase of disc temperatures (leading to dramatic increase of disc luminosity). The measured disc temperature versus luminosity relation suggests that the internal disc radius decreases [52]. Although other models such as dynamic accretion disc corona models cannot be ruled out, these results are consistent with the so-called truncated disc model [8].

Table 1. Best-fit parameters of the joint IBIS/ISGRI, JEM-X and SPI spectra (only rev 222, 223 and 224.1 for the spectrometer). Fits have been performed simultaneously with EQPAIR combined with DISKLINE. See Del Santo et al. (2008b) for detailed discussion.

Period	Rev	l_h/l_s	l_{nth}/l_h	τ_{es}	kT_{bb} [eV]	$\Omega/2\pi$	G_{inj}	χ^2_ν(dof)	Flux $\times 10^{-9}$ [erg cm^{-2} s^{-1}]		
									Bol	*bb*	*Compt*
1	222	$4.4^{+1.0}_{-0.4}$	$0.7^{+0.2}_{-0.1}$	$2.6^{+0.2}_{-0.9}$	(300)	$0.2^{+0.3}_{-0.1}$	$2.7^{+0.5}_{-0.6}$	0.99(243)	11.3	1.0	9.4
2	223	2.1 ± 0.4	$1.0^{+0.0}_{-0.2}$	$2.0^{+0.3}_{-0.5}$	(300)	0.6 ± 0.3	$2.6^{+0.3}_{-0.4}$	1.17(219)	12.4	2.1	8.9
3	224.1	$0.20^{+0.05}_{-0.07}$	$0.71^{+0.18}_{-0.09}$	$0.18^{+0.12}_{-0.03}$	380^{+54}_{-57}	$1.0^{+0.1}_{-0.3}$	$2.9^{+0.1}_{-0.2}$	1.14(215)	21.0	15.4	5.6
4	224.2	$0.17^{+0.03}_{-0.02}$	0.6 ± 0.1	$0.4^{+0.3}_{-0.1}$	388^{+76}_{-83}	(1)	$2.9^{+0.4}_{-0.2}$	1.06(173)	24.3	17.8	6.5
5	226	$0.15^{+0.13}_{-0.05}$	$0.94^{+0.06}_{-0.10}$	< 0.1	495^{+27}_{-43}	(1)	1.9 ± 0.3	1.15(121)	13.8	10.9	2.9
7	232-233	$0.07^{+0.04}_{-0.01}$	$1.0^{+0.0}_{-0.2}$	< 0.7	478^{+66}_{-83}	(1)	2.8 ± 0.5	1.0(98)	18.4	13.8	4.2

4.2. The high energy excess

During the Low/Hard state of the GX 339−4 outburst occurred in 1991, a high energy excess (above 200 keV) was observed with OSSE [54, 55]. In March 2004, i.e. during the increasing phase of the 2004 outburst, Joinet et al. (2007) observed a similar feature with SPI when GX 339−4 was in its canonical LH state.

The non-negligible values of l_{nth}/l_h (compactness ratio between the electron acceleration and total power supplied to the plasma) found also in the hard/intermediate state (spectra 1, 2 in Tab. 1) indicated a non-thermal emission being requested by the data. Indeed, using a simple thermal Comptonisation model, some residuals are present at high energy, especially in the spectrum collected during the revolution 223 [52]. Adding a power-law with $\Gamma \sim 2$ to the pure thermal Comptonisation model, it led to an improvement of the χ^2 (see [52] for a complete discussion).

Moreover, the SPI observations showed that during the bright hard state of the 2007 outburst, the highest energy emission (>150 keV) of GX 339−4 was variable [57]. While the spectral shape at lower energies (4-150 keV) remained more or less constant, Droulans et al. (2010) detected the significant appearance/disappearance of a high-energy tail. The strength of this hard tail, varying on a timescale of less than 7 hr, is found to be positively correlated with the total X-ray luminosity of the source. These authors interpreted the spectral variability as triggered by only the variations of the magnetic field in the hot electron plasma, implying a subsequent variation of the maximum energy of the accelerated particles [57].

5. The prototype of Galactic black hole binary: Cygnus X–1

Cygnus X–1, the archetype of the Galactic black holes discovered in 1964s [58], is a persistent X-ray binary with a supergiant star companion. Cyg X-1 is one of the best studied black hole binaries (BHBs) and it has served as an accretion disc laboratory since the end of the 1960s. In the LH state of Cyg X-1, the electron temperature and the Thomson optical depth of the Comptonising plasma were found to be typically $kT_e = 100$ keV and $\tau = 1$ [9]. Although the scattering plasma is predominantly thermal in the hard state, since 1994 there are indications that the electron distribution has some high-energy tail i.e., it is hybrid, thermal/nonthermal. This evidence was firstly provided by *CGRO*/COMPTEL observations of Cyg X-1 in the hard state at $E > 500$ keV[59]. The X/γ-ray HS state spectrum has been studied extensively by simultaneous observations with *ASCA*, *RXTE*, *BeppoSAX*, and *CGRO* during summer 1996[60, 61]. In addition to the dominating black-body, a long power-law like tail extending up to 10 MeV was discovered [62].

As one of the brighter galactic hard X-ray source, Cyg X-1 is a prime target for the *INTEGRAL* mission. It was extensively observed during the Performance Verification (PV)

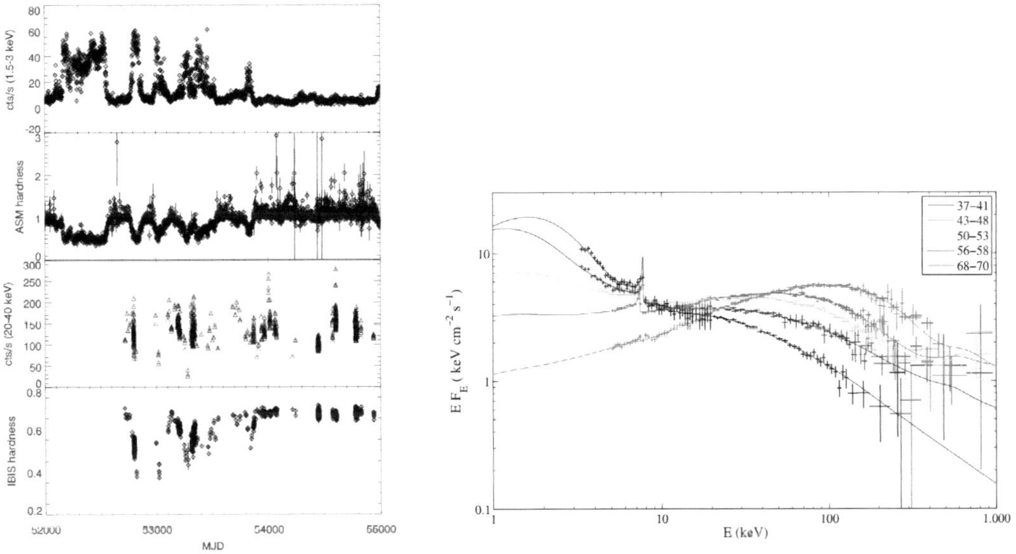

Figure 7. *Left*: From top to bottom: ASM count rate in 1.5-3 keV, ASM hardness ratio (3-6/1.5-3 keV), IBIS/ISGRI count rate in 20-40 keV, IBIS/ISGRI hardness (40-80/20-40 keV). *Right*: Joint JEM-X, IBIS and SPI energy spectra of Cyg X-1 collected in different spectral states. The data are fitted with model EQPAIR plus DISKLINE. Spectral parameters will be presented in Del Santo et al. (2012).

Phase of the mission, when the source was in the HSS[63, 64, 65]. The on-board instruments offer an unprecedented simultaneous broad-band spectral coverage, ranging from 3 keV to several MeV. Thus, up to now a big amount of observing time (open time and core programme for a total of about 7 Ms) have been dedicated to this target by *INTEGRAL*. A number of studies of the intermediate spectral state have been performed [66, 68]; a small X-ray flare during the hard state have been observed to be coincident with the TeV emission detected by MAGIC [67]. Recently, Laurent et al. (2011) used the IBIS telescope in Compton mode and measured the polarization of the gamma-ray emission from Cyg X-1. While the 250-400 keV spectrum is consistent with emission dominated by Compton scattering by thermal electrons and are weakly polarized (Pf $<20\%$), the second spectral component seen in the 400 keV–2 MeV band is strongly polarized (Pf $= 67\pm 30\%$). The authors proposed that the MeV emission is probably related to the jet detected in the radio band.

However, Jourdain et al. (2001)report on Cyg X-1 observations performed by the SPI telescope distributed over more than 6 years. They investigate the hard X-rays domain, and more particularly up to the MeV region. They found that the Cyg X-1 emission decreases sharply above 700 keV, with flux values above 1 MeV (or upper limits) well below to the one reported in[69], while compatible with the MeV emission detected some years ago by *CGRO*/COMPTEL[62].

5.1. Long-Term Spectral Variability of Cygnus X–1 with INTEGRAL

We have analysed six years of Cygnus X–1 observations performed by *INTEGRAL* since March 2003 until April 2009 with the three telescopes IBIS, SPI and JEM-X. In Fig. 7 (*left*), panels 3 and 4 (from top to bottom), we show the IBIS light curve and the hardness ratio of Cyg X-1 in the hard X-ray domain. As a comparison at the soft X-rays, we plot the *RXTE*/ASM long term count rate, as well as the hardness ratio (Fig. 7 (*left*), panels 1 and 2). During the first three years of the *INTEGRAL* monitoring, Cyg X-1 showed a very variable activity in term of both flux variation and spectral transition. Since middle of 2006, the source entered in an almost steady hard state, characterized by flux variability not combined with simultaneous spectral variability (Fig. 7, *left*). The work is in progress and will be presented in Del Santo et al. (2012). In that paper, we will report on the long term behaviour of Cyg X-1 using the whole *INTEGRAL* data base available until 2009. A broad-band spectral variability study by using two different models over a wide energy band (from 3 keV up to 1 MeV) spectra has been performed. In Fig. 7 (*right*) I show preliminary Cyg X-1 energy spectra fitted by the hybrid thermal/non-thermal Comptonisation plus a disc reflection model, namely EQPAIR [53], and a relativistic iron line line emission.

6. Acknowledgments

I acknowledge the following funding: the agreement ASI-INAF I/009/10/0, PRIN-INAF 2009 (PI: L. Sidoli), IASF-Roma and Université Toulouse III. I acknowledge a big number of collaborators, among them J. Malzac (CNRS/IRAP). This work is based on observations with INTEGRAL, an ESA project with instruments and Science Data Centre funded by ESA member states (especially the PI countries: Denmark, France, Germany, Italy, Switzerland, Spain), Czech Republic, and Poland, and with the participation of Russia and the USA.

7. References

[1] Frank J, King A, and Derek R 2002 Accretion Power in Astrophysics, ed. Cambridge
[2] Shakura N. I., & Sunyaev R. A., 1973, A&A 24, 337
[3] Balbus S. A., & Hawley J. F., 1991, ApJ, 376, 214
[4] Balbus S. A., & Hawley J. F., 1998, Rev. Mod. Phys. 70, 1
[5] Narayan R., & Yi I., 1994, ApJ, 428, L13
[6] Psaltis D., 2008, "Compact stellar X-ray sources". Ed. W. Lewin & M. van der Klis. Cambridge Astrophysics Series, p. 1
[7] Zdziarski A. A., 2000, IAU Symp.,195, 153
[8] Done C., Gierliński M., Kubota A., 2007, A&A Review, 15, 1
[9] Zdziarski A. A. & Gierliński M., 2004, Progress of theoretical Physics, 155, 99
[10] Tanaka Y., & Lewin W.H.G., 1995, in "X–ray Binaries", eds. W.H.G. Lewin, J. van Paradijs and E.P.J. van den Heuvel, Cambridge U. Press, Cambridge)
[11] McClintock J. E. & Remillard R. A., 2008, "Compact stellar X-ray sources". Ed. W. Lewin & M. van der Klis. Cambridge Astrophysics Series, p. 157
[12] Rhoades C.E. & Ruffini, R., 1974, Phys.Rev.Lett., 32, 324
[13] Kalogera V., & Baym G., 1996, ApJ, 470, L61
[14] Capitanio F., et al., 2009, MNRAS, 398.1194C
[15] Belloni T., et al., 2005, A&A, 440, 207
[16] Campana, S., Parmar, A. N., Stella, L., 2001, A&A, 372, 241.
[17] Homan J., Belloni T., 2005, ApSS, 300, 107
[18] Fender R.P., Belloni T.M., Gallo E., 2004, MNRAS, 355, 1105
[19] Winkler C., et al., 2003, A&A, 411, L1
[20] Ubertini P., et al. 2003, A&A, 411, L131
[21] Vedrenne G,. et al. 2003 A&A, 411, L63
[22] Lund N., et al. 2003, A&A, 411, L231
[23] Mass-Hesse M. et al. 2003, A&A, 411 L261
[24] Skinner G., & P. Connell, 2003, A&A, 411, L123
[25] Courvoisier T. J. L., et al., 2003, A&A, 411, L53
[26] Lebrun, F., et al. 2003, A&A, 411, L141
[27] Bird A. J., et al., 2004, ApJ, 607, L33
[28] Bird A. J., et al., 2006, A&A, 445, 869
[29] Bird A. J., et al., 2007, ApJS, 170, 175
[30] Bird A. J., et al., 2010, ApJS, 186, 1
[31] Hertz P. & Grindlay J. E., 1984, ApJ, 278, 137
[32] Sunyaev R. et al. 1991, ApJ, 383, L49

[33] Mirabel F., Rodriguez L. F., Cordier B., Paul J. & Lebrun F., 1992, Nature, 358, 215
[34] Smith D. M., Heindl W. A. & Swank J. H. 2002, ApJ, 569, 362
[35] Del Santo M., et al., 2005, A&A, 433, 613
[36] Del Santo M., et al., 2008a, Proceed. of INTEGRAL Workshop, PoS, E. 100
[37] Bouchet L., Del Santo M., Jourdain E., Roques J. P., Bazzano A., De Cesare G., 2009, 693, 1871
[38] Titarchuk L., 1994, ApJ, 434, 313
[39] McConnell M. L., et al. 2000, ApJ, 543, 928
[40] Zdziarski A. A., Grove E., Poutanen J., Rao A. R., & Vadawale S. V., 2001, ApJL, 554, L45
[41] Malzac J., & Jourdain E., 2000, A&A, 359, 843
[42] Bouchet L., et al., 1991, ApJ, 383, L45
[43] Sunyaev R., et al., 1991, ApJ, 383, L49
[44] Cheng L. X., et al., 1998, ApJ, 503, 809
[45] Teegarden B. J. & Watanabe 2006, ApJ, 646, 965
[46] De Cesare G., 2011, A&A, 531, 56
[47] van Oss R. F. & Belyanin A. A., 1991, A&A., 302, 154
[48] Markert T. H., Canizares C. R., Clark G. W., et al., 1973, ApJ, 184, L67
[49] Shahbaz T., Fender R. P., Charles P. A., 2001, A&A, 376, L17
[50] Zdziarski A. A., et al., 1998, MNRAS, 301, 435
[51] Zdziarski A. A., et al., 2004, MNRAS, 351, 791
[52] Del Santo M., et al., 2008b, MNRAS, 390, 227
[53] Coppi P. S., 1999, ASP Conf. Ser. 161: High Energy Processes in Accreting Black Holes, 161, 375
[54] Johnson W. N., et al., 1993, A&AS, 97, 21
[55] Wardziński G., et al., 2002, MNRAS, 337, 829
[56] Joinet A., et al., 2007, ApJ, 657, 400
[57] Droulans R., et al., 2010, ApJ, 717, 1022
[58] Bowyer S., Byram E. T., Chubb T. A., & Friedman M., 1965, Science, 147, 394
[59] McConnell M. L., et al., 1994, ApJ, 424, 933
[60] Di Salvo T., et al., 2001, ApJ, 547, 1024
[61] Frontera F. et al., 2001, ApJ, 546, 1027
[62] McConnell M. L. , et al., 2002, ApJ, 572, 984
[63] Bazzano A., et al., 2003, A&A, 411, L389
[64] Bouchet L., et al., 2003, A&A, 411, L377
[65] Pottschmidt K., et al., 2003, A&A, 411, L383
[66] Malzac J., et al., 2006, A&A, 448, 1125
[67] Malzac J., et al., 2008, A&A, 492, 527
[68] Cadolle Bel M., et al., 2006, A&A 446, 591
[69] Laurent P., et al., 2011, Science, 332, L438
[70] Jourdain E., Roques J.-P., Malzac J., 2011, ApJ accepted (astro-ph/1109.2053)

3rd Italian–Pakistani Workshop on Relativistic Astrophysics (IPWRA2011) IOP Publishing
Journal of Physics: Conference Series **354** (2012) 012004 doi:10.1088/1742-6596/354/1/012004

CMB as a possible new tool to study the dark baryons in galaxies

F. De Paolis, G. Ingrosso, A.A. Nucita, D. Vetrugno
Dipartimento di Fisica, Università del Salento, Via per Arnesano, I-73100, Lecce, Italy and INFN, Sezione di Lecce, Via per Arnesano, I-73100, Lecce, Italy

E-mail: francesco.depaolis@le.infn.it

V.G. Gurzadyan, A.L. Kashin, H.G. Khachatryan, S. Mirzoyan
Yerevan Physics Institute and Yerevan State University, Yerevan, Armenia

Ph. Jetzer
Institute für Theoretische Physik, Universität Zürich, Winterthurerstrasse 190, 8057 Zürich, Switzerland

A. Qadir
Center for Advanced Mathematics and Physics, National University of Sciences and Technology, H-12, Islamabad, Pakistan

Abstract. Baryons constitute about 4% of our universe, but most of them are missing and we do not know where and in what form they are hidden. This constitute the so-called missing baryon problem. A possibility is that part of these baryons are hidden in galactic halos. We show how the 7-year data obtained by the WMAP satellite may be used to trace the halo of the nearby giant spiral galaxy M31. We detect a temperature asymmetry in the M31 halo along the rotation direction up to about 120 kpc. This could be the first detection of a galactic halo in microwaves and may open a new way to probe hidden baryons in these relatively less studied galactic objects using high accuracy CMB measurements.

1. Introduction

We have entered in a new era of high precision cosmology and it is clear by primordial nucleosynthesis constraints and by the results from the power spectrum of the CMB produced by the 7-years data of WMAP satellite that our universe is composed of $\simeq 73\%$ of dark energy (DE), $\simeq 23\%$ of non-baryonic dark matter (DM) and for only $\simeq 4\%$ by baryons. About baryons, we know that about 10% of them are in stars (i.e. in the visible part of galaxies); the hot gas in galaxies and galaxy clusters accounts for another $20 - 30\%$ of baryons; but about $60 - 70\%$ of all baryons are missing and is not known where and in which form they are hidden [1]. A possibility is that these baryons are contained in the so-called cosmic filaments in the form of a

[1] Observations also show that galaxies are missing most of their baryons with respect to the cosmological baryon/matter ratio (see e.g. [1, 2, 3]).

warm-hot intergalactic medium (see e.g. [4, 5, 6, 7]). However, it is unlikely that all the missing baryons are confined in the cosmic filaments and, actually, there are many reasons to believe that a non-negligible amount of these hidden baryons are in galactic halos (see below in this section) or even in the galactic disk (see e.g. [8] and references therein) [2].

A rather clear piece of evidence for the existence of large amounts of gas around galaxies is the widespread detection of absorption lines in front of quasars, the so-called Ly-α forest, that show hundreds of absorption systems along a single line of sight. Also damped Ly-α systems with hydrogen columns density larger than about 10^{20} cm^{-2} are associated with galaxies and both systems imply that galaxies at high redshift ($\sim 2-2.5$) would have huge gas halos around them. Where is all that gas now? One possibility is that it has been expelled from galaxies during the first chaotic phase of star formation into the cosmic filaments. However, that might have happened only in part, leaving nowadays some of that gas around galactic disks.

There is further evidence that there is some amount of gas in galactic halos, and in particular in that of our galaxy: the problem is to measure how much it is. That is, obviously, not so easy. Let us briefly discuss some of this evidence.

Galaxies seem to evolve along the Hubble sequence from Sc type, characterized by small bulges and open spiral arms to Sa, characterized by larger bulges and tighter spirals. During this evolution the M/L ratio, which is a measure of the amount of dark matter, decreases (while the number of stars increases) implying that part of the dark matter should transform into stars and that means that part of the dark matter should consist of baryons in gaseous form. Moreover, star-forming disk galaxies similar to our Milky Way, should exhaust their cold gas reservoir (necessary to make stars) within a few Gyrs unless it is replenished in some way [10, 11]. Numerical simulations (see e.g. [12]) indeed show that galaxies like the Milky Way convert about 1 $M_\odot$ yr^{-1} of gas into stars while their gas content remains approximately unchanged. The reservoir of the accretion material is as yet unidentified but it goes without saying that it should come from the region that surrounds the galactic disks and might even reside in the intergalactic medium (see [13] for a discussion about the cooling mechanisms of the hot intergalactic gas). In any case, a gas accretion rate of about 1 $M_\odot$ yr^{-1} for M31/Milky Way galaxies onto their galactic disks is compulsory (see also [14]).

The wealth of data in the last two decades show that there is good evidence for extra-planar gas around the Milky Way and other galaxies and it is detected in all gaseous phases: neutral, warm atomic, molecular and hot X-ray emitting gas. Since about a decade we know that around the Milky Way and Andromeda (M31) galaxies there is a population of high-velocity clouds (HVCs) that extends up to $10-20$ kpc and are not seen beyond about 50 kpc [15]. There are also intermediate-velocity clouds (IVCs) with a vertical scale height of $4-5$ kpc and a total mass about $3-4\times 10^8$ $M_\odot$, forming a population separated with respect to that of the HVCs. The innermost halo clouds show a metallicity that is about half of the solar one and this favors the Galactic fountain model for its origin in which hot gas is ejected out of the galactic disk, cools down and falls back onto the disk (see e.g. [16]). As far as HVCs are concerned, since generally only the radial velocity can be estimated for these objects and distance information are sparse and not well known, a closed theory about their origin is not available today. One possibility is that they represent metal poor gas from the intergalactic space falling onto the galactic disk [17], the other is that they represent the remnant of the galactic halo formation.

Also around the Andromeda galaxy (M31) deep radio synthesis observations has shown the presence of gas clouds within about 50 kpc. These clouds, which are the analogue of the HVCs, have size of about 1 kpc and mass $\simeq 10^5$ $M_\odot$ [18]. The total mass in HVCs is about 10^8 $M_\odot$ [14].

[2] The first to suggest the concept of the presence of the gas above the disk of the Milky Way was Spitzer in his remarkable paper of 1956 [9] where he also discussed the nature and the possible evolution of the gas eventually present there.

It also seems that a dilute gas halo at a temperature $\sim 10^6$ K is present around some spiral galaxies. Elliptical galaxies (and particularly bright ellipticals), as is well known, do contain large amount of hot diffuse gas emitting in X-rays (see e.g. [19] and references therein). For spirals, evidence is much less convincing. A difficulty in this respect is that the emission measure scales with the square of the electron density and once the galaxy surface brightness drops to the level of the X-ray background (mainly due to the Milky Way) further detection is impossible. That is the reason why the size of the X-ray emitting gas around spiral galaxies is very poorly constrained. Interestingly enough, very recently a hot gaseous halo has been detected by XMM-Newton satellite towards UGC12591, the fastest rotating spiral galaxy [20], up to a distance of about 110 kpc from the galaxy center. Combining the X-ray data with near-IR and radio measurements it has been found that the baryon mass fraction in this galaxy is about $3-4\%$ (in particular the baryon mass within about 500 kpc is $\simeq 5.9\times10^{11}\ M_{\odot}$ while the total mass, estimated using the galaxy rotation curve, is found to be $\simeq 2.7\times10^{13}\ M_{\odot}$). This would imply that the majority of the missing baryons in spiral galaxies does not reside in their hot gas halos [3].

Indeed, after the detection of the first microlensing events towards the LMC [22, 23] [4] a model was proposed for the formation of MACHOs (Massive Astrophysical Compact Halo Objects) in the galactic halo [44, 45]. Actually, this model naturally emerges from the present-day understanding of the globular cluster formation. Indeed, the Fall-Rees theory for the formation of globular clusters [46] predicts, without any further assumption, that dark clusters made of brown dwarfs and cold gas clouds should lurk in the galactic halo at galactocentric distances larger that $10-20$ kpc. Accordingly, the inner halo is populated mainly by globular clusters, whereas the outer galactic halo should be dominated by dark clusters populated by MACHOs and cold gas clouds [5]. The gas clouds in the dark clusters should be very cold, with a temperature very close to that of the CMB, should have sub-solar mass and size about that of the Solar System (for details see [53]).

A novel way of investigating the amount and distribution of the dark baryons in the galactic halos, and in particular in that around the M31 galaxy is discussed in the next section. This method deals with the use of CMB data (at present the 7-year WMAP data) to trace galactic disks and halos.

2. The temperature asymmetry in the M31 halo by the 7-year WMAP data

Galactic disks are well studied objects in all wavelengths and give important information on the mass distribution within and around galaxies [54]. On the other hand, galactic halos are relatively less studied structures and there are still many ambiguities not only in the main halo constituents, but also in the basic properties such as, in particular, the rotation.

The degree to which galactic halos rotates with respect to the disks is a relevant and difficult

[3] The fact that galactic halos do not contain much hot baryons is also confirmed by the dispersion measure of pulsars in the LMC [21].

[4] Microlensing observations conducted since two decades both towards the Galactic bulge and the Magellanic Clouds by MACHO, EROS and OGLE Collaborations have shown that a (probably) not-negligible fraction (about $5-20\%$) of the halo dark matter may be made by MACHOs (see e.g. [24, 25, 26]) although this result as well as the estimation of the average MACHO mass implied by observations depends on the specific adopted halo model (see e.g. [27, 28]). Pixel-lensing observations towards the M31 galaxy have been conducted by several collaborations leading to the discovery of more than 35 events [29, 30, 31, 32, 33, 34, 35] up to now (one of which even showed the presence of an exoplanet in Andromeda galaxy [36, 37]) give uncertain conclusions about the fraction of the halo dark matter in the form of MACHOs, ranging from about 20% for $0.1-1\ M_{\odot}$ MACHOs to the possibility of explaining the detected events simply by self-lensing (see e.g. [38, 39, 40, 41, 42, 43]).

[5] Somewhat similar scenarios have been also proposed in [47, 48, 49, 50]. Models in which the galactic dark matter is in baryonic form and distributed in a thin disk have also been proposed along with many observational consequences [51, 52].

issue to be investigated. The rotation of the galactic halos, indeed, is clearly related to the formation scenario of galaxies. In the standard collapse model (see e.g. [55]) both the halo and disk of galaxies derive from the same population. The rotation of the outer halo should be, in this case, aligned with the disk angular momentum. On the contrary, in a hierarchical formation scenario, structures arriving later in the outer halo should have a minor connection to the disk. Therefore, it is evident that information on the galactic halo rotation provide key insights about the formation history of galaxies. Nevertheless, it is important to stress that testing for the rotation even of the closest galaxy (M31) halo is still beyond our reach [56] (see also [57]).

The first attempt of using the 7-year WMAP data [58] in the three bands W (94 GHz), V (61 GHz), and Q (41 GHz) to map in microwaves both the disk and halo of the M31 galaxy is provided in [59] [6]. To reveal the different contributions by the M31 disk and halo, the region of the sky towards the M31 galaxy has been divided into several concentric circular areas as shown in Fig. 1 in [59].

The M31 disk does contain gas observed mainly at 21 cm wavelength (but also in the IR) up to a distance of about 40 kpc from the galactic center. It is also well known that the M31 disk rotate with a speed of about 250 km s^{-1} and this has been clearly shown also by the velocity maps provided in [60, 61]. In [59] a temperature asymmetry along the direction of the M31 rotation has been observed for the first time also in microwaves with a maximum of $\simeq 130$ μk/pixel at about 20 kpc from the M31 center. This temperature asymmetry is very likely induced by the Doppler shift effect due to the M31 disk rotation speed. The robustness of this result has been tested by considering 500 randomly distributed control fields in the three WMAP bands and also by simulating 500 sky maps from the best fit cosmological parameters. Both procedures give comparable results and imply that there is less than $\simeq 2\%$ probability that the signal is due to a random fluctuation of the CMB signal. An analogous study has been also conducted in the same paper towards the M31 halo within 20^0 (about 240 kpc from the M31 center). We found also towards the M31 halo a CMB temperature asymmetry up to about 120 kpc with a peak temperature contrast of about 40 μK/pixel. Although the confidence level of the signal, if estimated purely statistically (i.e. with 500 control fields and 500 simulated sky maps), is not high (we find, indeed, that there is a probability of less than about 30% that the detected temperature asymmetry in the M31 halo is due to a random fluctuation of the CMB signal) [7], the geometrical structure of the temperature asymmetry in the three bands point towards a real effect modulated by the rotation of the M31 halo.

We point out once more that the use of three WMAP bands is important for revealing the role of the contribution of the Galactic foregrounds since each emission mechanism contributes differently in each band. The fact that the temperature contrast seems present in all three bands and is more or less the same in each band up to about $10^0 - 11^0$ (about 120 kpc from the M31 center) indicates that the foregrounds are far weaker than the effect. A size of about 120 kpc corresponds to the typical size inferred for the dark matter halos around massive galaxies and might open the possibility of a new way of studying these systems, both galactic disks and halos, at microwave wavelengths. In any case, a careful analysis of the Planck data that should be released shortly should allow either to prove or disprove our main results.

[6] The use of three WMAP bands is important in revealing the possible contribution of the Galactic foregrounds since dust, free-free, and synchrotron emission contributes in a different way in each band. We also remind that the band least contaminated by the synchrotron radiation of the Galaxy is the W-band, which also has the highest angular resolution.

[7] Actually, if one takes the direction of rotation of the M31 disk into account, such a probability reduces (by using the theorem of the composite probability) by a factor of two. So that one can conclude that there is a probability of less than about 15% that the detected temperature asymmetry in the M31 halo is due to a random fluctuation of the CMB signal.

3. Discussion

As clear from the discussion above, the detected temperature asymmetry for the M31 disk is fairly clear in all WMAP bands, and is also expected due to the foreground emission of the M31 disk modulated by the Doppler shift induced by the disk rotation. Incidentally, the M31 galaxy has been recently detected by the Planck observatory [62] [8], whereas it did not appear in the WMAP list. These are all reasons to expect that the particular effect we discuss here can be studied more accurately with Planck data.

As for the M31 halo, we have shown that, although less evident than for the M31 disk, there is less than about 20% probability that the detected temperature asymmetry at a galactocentric distance ~ 50 kpc comes from a random fluctuation of the CMB signal. [9]

If one assumes that this temperature asymmetry in the M31 halo relies in the M31 itself and is related to the M31 halo rotation, a natural question that arises is about the origin of this effect. In all generality, four possibilities may be considered: (*i*) free-free emission; (*ii*) synchrotron emission; (*iii*) Sunyaev-Zel'dovich (SZ) effect; and (*iv*) cold gas clouds populating the M31 halo. The first three effects assume the presence of a rather hot plasma in the halo of M31. Although this hot plasma has not been detected yet, one can assume that a certain amount of this plasma can populate the M31 halo (spiral galaxies are believed to have much less hot gas than ellipticals) and may rotate with a certain speed. Free-free emission arises from electron-ion scattering while synchrotron emission comes mostly from the acceleration of cosmic-ray electrons in magnetic fields. Both effects give rise to a thermal emission with a rather steep dependence on the frequency [64] that therefore should give a rather different temperature contrast in the three WMAP bands. The absence of this effect indicates that the contribution from possibilities (*i*) and (*ii*) should be negligible. In the case of (*iii*), even for typical galaxy clusters with diffuse gas much hotter than that possibly expected in the M31 halo, the rotational effect produces a temperature asymmetry of at most a few μK pixel^{-1}, depending on the rotational velocity and the inclination angle of the rotation axis [65], that is much less than that observed towards the M31 halo. Actually, a possible temperature asymmetry in the CMB data towards the M31 halo as a consequence of the existence of a population of cold gas clouds in its halo was predicted in [66] - possibility (*iv*). Indeed, if the halo of the M31 galaxy would contain gas clouds, one expects them to rotate along the disk rotation (even if, perhaps, more slowly), and thus there should be a Doppler shift inducing a temperature anisotropy ΔT between one side of the M31 halo and the other with respect to the rotation axis perpendicular to the disk. In the case of optically thin halo clouds the Doppler induced temperature asymmetry would be $\Delta T/T_r \simeq 2vS\bar{\tau}/c$, where $T_r \simeq 2.725$ K is the CMB temperature, v is the M31 averaged rotation speed, $\bar{\tau}$ the averaged cloud optical depth over the frequency range ($\nu_1 \leq \nu \leq \nu_2$) of a certain detection band, and S the cloud filling factor, i.e. the ratio of filled (by clouds) to total projected sky surface in a given field of view. In the case of optically thick gas clouds instead one has $\Delta T/T_c \simeq 2vS/c$, where T_c is the cloud temperature [10].

In [66] we concentrated mainly on the optically thick option and, as a suggested test for the proposed model (see also [67, 68]), made a rough estimate [11] of the expected temperature asymmetry between the two opposite sides of the M31 halo as $\Delta T/T_r \simeq 2 \times 10^{-5}\bar{\tau}$ that implies

[8] However, there is no mention of any temperature asymmetry in the M31 disk in that paper.

[9] We also mention that the number and the temperature profile of radio sources in CMB maps [63] excludes their significant contribution in the effect under study.

[10] Note, however, that cold gas clouds in galactic halos are not expected to be optically thick at microwave wavelengths.

[11] We estimated, based on our model, S to be about 1/25 and assumed a halo rotation speed about 100 km s^{-1}. By the way, this value for S is not far from that coming from recent observations of HVCs in M31 presented in Fig. 1 in [14]. Indeed it turns out that S can be parameterized as $S \simeq 2.1\exp[-r/(12kpc)]$, so that $S \simeq 0.05$ at about 40 kpc from the M31 center. We emphasize, however, that this limit has to be considered as a lower limit to the true value of S since it only comes from considering the observed HVCs in the M31 halo.

a temperature asymmetry about 50 μK for $\bar{\tau} \lesssim 1$, that is very close to the value recently found in [59].

Before closing this Section we would like to mention that one of the most serious problems associated with the presence of cold gas clouds in galactic halos is that of explaining their mere dynamical stability. As a matter of fact, isothermal perfect gas clouds without fixed boundaries are unstable with respect to the gravothermal catastrophe on time scales of a few crossing times. Generally speaking, the outcome of the gravitational collapse of a gas cloud is the formation of a dense central core (see e.g. [69]). At the beginning, the evolution is almost isothermal (due to the low optical depth of the gas cloud) but, when opacity exceeds unity, the central temperature and pressure rapidly increase, leading to star formation. However, halo cold gas clouds may mostly succeed in avoiding collapse since in low temperature and low opacity conditions, collapse may lead to the formation of solid/liquid H_2 (below about 14 K) and to the formation of a temperature inverted gas cloud with a condensed cold core. This new physical ingredient provides the possibility of stabilizing these cold gas clouds. Such loosely bound cold clouds could form a gas reservoir in galaxy halos and behaves as a collisionless ensemble of matter. Incidentally, these cold cores might also evaporate due to ambient heating, but they can leave for Gyrs in the clod and low excitation environment such as those expected in galactic halos (for further details see also [70, 71]). Interestingly enough, cosmic rays and energetic photons (for example those in the UV band of the electromagnetic spectrum) might cause ionization in the solid molecular hydrogen leading to the formation of H_6^+ and $(HD)_3^+$. Very recently, it has been shown [72] that the so-called unidentified infrared (UIR) bands, usually thought to be related to large molecules such as polycyclic aromatic hydrocarbons, may instead be produced by $(HD)_3^+$ and H_6^+ in solid form, thus showing that solid H_2 may indeed be abundant in quiet astrophysical environments and opening a new way of searching for cold gas clouds.

4. Concluding remarks

As mentioned earlier, it is likely that there are no grounds to assume that a unique solution may solve the hidden baryon problem. It is instead plausible that it is a composite problem reflecting the diversity of astrophysical conditions.

For example, globular clusters are mostly composed of baryons and no evidence for non-baryonic mass is present in these objects. Thus, baryons are well separated from non-baryonic dark matter in the formation process of these structures. On the galaxy cluster scale most baryons are in the form of a hot X-ray emitting gas while in the inter-galactic medium most baryons might be in the form of a warm-hot gas.

The wealth of data especially in the last decade shows that there is good evidence for the presence in the halos of spiral galaxies of gas in all gaseous phases: neutral, warm atomic, and hot X-ray emitting gas [73]. Atomic gas (often identified as HVCs) is observed in the radio band (particularly at 21 cm) and through absorption lines towards field stars and quasars. The hot gas may be detected in X-rays but, unfortunately, searches for cold gas clouds in galactic halos are more problematic. Various attempts have been undertaken to detect such clouds as searching for the presence of a gamma-ray halo [74, 75], stellar scintillations [76, 77], obscuration events towards the LMC [78], ortho-H_2D^+ line at 372 GHz [79], and extreme scattering events in quasar radio-flux variations [50]. All these searches have given no clear indication of cold gas cloud presence in galactic halos, but these searches are going on using more and more sensitive facilities and new developments, such as searching for the vibrational transition lines produced by $(HD)_3^+$ and H_6^+ in solid form. Therefore, the issue of a realistic estimate of the amount of baryonic mass in galactic halos and their size does remain open and a new perspective in this research can be given by investigating it in microwaves following [59].

Acknowledgments
VG and AQ acknowledge some financial support by the Department of Physics, University of Salento, and by INFN.

References

[1] Corbelli E 2003 Month. Not. R. Astron. Soc. **342** 199
[2] Hoekstra H *et al* 2005 Astrophys. J. **635** 73
[3] Bregman J N 2007 Ann. Rev. Astron. Astrophys. **45** 221
[4] Cen R and Ostriker J P 1999 Astrophys. J. **514** 1
[5] Cen R and Ostriker J P 2006 Astrophys. J. **650** 560
[6] Fraser-McKelvie A *et al* 2011 Month. Not. R. Astron. Soc. **415** 1961
[7] Nicastro F, Mathur S and Elvis M 2008 Science **315** 55
[8] Revaz Y *et al* 2009 Astron. Astrophys. **501** 171
[9] Spitzer L 1956 Astrophys. J. **124** 20
[10] Chiappini C, Matteucci F and Romano D 2001 Astrophys. J. **554** 1044
[11] Sommer-Larsen J, Götz M and Portinari L 2003 Astrophys. J. **596** 47
[12] Twarog B A 1980 Astrophys. J. **242** 242
[13] Binney J *et al* Month. Not. R. Astron. Soc. **397** 1804
[14] Richter P 2011 Preprint arXiv:1111/1717v1
[15] Wakker B P *et al* 2008 Astrophys. J. **672** 298
[16] Richter P *et al* 1999 Nature **402** 386
[17] Oort J H 1970 Astron. Astrophys. **7** 381
[18] Westmeier T *et al* 2007 New Astronomy Review **51** 108
[19] De Paolis F, Ingrosso G and Strafella F 1995 Astrophys. J. **438** 83
[20] Dai X *et al* 2011 Preprint arXiv:1112.0324v1
[21] Anderson M E and Bregman J N 2010 Astrophys. J. **714** 320
[22] Aubourg E *et al* 1993 Nature **365** 623
[23] Alcock C *et al* 1993 Nature **365** 621
[24] Alcock C *et al* 2000 Astrophys. J. **542** 281
[25] Tisserand P *et al* 2007 Astron. Astrophys. **469** 387
[26] Calchi Novati 2007 Nuovo Cimento B **122** 557
[27] De Paolis F, Ingrosso G and Jetzer Ph 1996 Astrophys. J. **470** 493
[28] Grenacher L *et al* 1999 Astron. Astrophys. **351** 775
[29] Ansari R *et al* 1999 Astron. Astrophys. Lett. **344** L49
[30] Aurière M *et al* 2001 Astrophys. J. **553** L137
[31] Calchi Novati S *et al* 2002 Astron. Astrophys. **381** 848
[32] Riffeser A *et al* 2003 Astrophys. J. **599** L17
[33] de Jong J T A *et al* 2004 Atron. Astrophys. **417** 461
[34] Calchi Novati S *et al* 2007 Astron. Astrophys. **469** 115
[35] Calchi Novati S *et al* 2009 Astrophys. J. **695** 442
[36] Ingrosso G *et al* 2009 Month. Not. R. Astron. Soc. **399** 219
[37] Ingrosso G *et al* 2011 Gen. Relat. Gravit. **43**1047
[38] Calchi Novati S *et al* 2005 Astron. Astrophys. **443** 911
[39] De Paolis F *et al* 2005 Astron. Astrophys. **432** 501
[40] de Jong J T A *et al* 2006 Astron. Astrophys. **446** 855
[41] Ingrosso G *et al* 2006 Astron. Astrophys. **445** 375
[42] Ingrosso G *et al* 2007 Astron. Astrophys. **462** 895
[43] Calchi Novati S 2010 Gen. Relat. Gravit. **42** 2101
[44] De Paolis F *et al* 1995 Phys. Rev. Lett. **74** 14
[45] De Paolis F *et al* 1995 Astron. Astrophys. **295** 567
[46] Fall S and Rees M 1985 Astrophys. J. **298** 18
[47] Ashman K M 1990 Month. Not. R. Astron. Soc. **247** 662
[48] Gerhard O and Silk J 1996 Astrophys. J. **472** 34
[49] Draine B 1998 Astrophys. J. Lett. **509** L41
[50] Walker M and Wardle M 1998 Astrophys. J. Lett. **498** L125
[51] Pfenniger D Combes F and Martinet L 1994 Astron. Astrophys. **285** 79
[52] Sciama D 2000 Month. Not. R. Astron. Soc. **312** 33
[53] De Paolis F *et al* 1998 Astrophys. J. **500** 59

[54] Binney J and Merrifield M 1998 Galactic Astronomy (Princeton Series in Astrophysics)
[55] Eggen O J, Lynden-Bell D and Sandage A R 1962 Astrophys. J. **136** 748
[56] Courteau S *et al* 2011 Astrophys. J. **739** 20
[57] Deason A J, Belokurov V and Evans N W 2011 Month. Not. R. Astron. Soc. **411** 1480
[58] Jarosik N *et al* 2011 Astrophys. J. Suppl. **192** 14
[59] De Paolis F *et al* 2011 Astron. Astrophys. Lett. **534** L8
[60] Chemin L, Carignan C and Foster T 2009 Astrophys. J. **705** 1395
[61] Corbelli E *et al* 2010 Astron. Astrophys. **511** A89
[62] Ade P A R *et al* (Planck Collaboration) 2011 arxiv:1101.2045v1
[63] Gurzadyan V G *et al* 2010 Europhys. Lett. **91** 19001
[64] Bennett C L 2003 Astrophys. J. **148** 97
[65] Cooray A and Chen X 2002 Astrophys. J. **573** 43
[66] De Paolis F *et al* 1995 Astron. Astrophys. **299** 647
[67] De Paolis F *et al* 1996 Astrophsy. Sp. Sci. **235** 329
[68] De Paolis F *et al* 1996 Int. J. Mod. Phys. D **5** 151
[69] Larson R B 1969 Month. Not. R. Astron. Soc. **145** 271
[70] Pfenniger D and Combes F 1994 Astron. Astrophys. **285** 93
[71] Pfenniger D 2004 in Dark Matter in Galaxies Symp. (Ryde S D et al. eds.), IAU Proceedings **220** 241
[72] Lin C Y, Gilbert A T B and Walker M A 2011 Astrophys. J. **736** 91L
[73] Bregman J N 2009 Preprint arxiv:0907.3494v1
[74] Dixon D D *et al* 1998 New Astronomy **3** 539
[75] De Paolis F *et al* 1999 Astrophys. J. Lett. **510** L103
[76] Moniez M 2003 Astron. Astrophys. **412** 105
[77] Habibi F *et al* 2011 Astron. Astrophys. **525** 108
[78] Drake A J and Cook K H 2003 Astrophys. J. **589** 281
[79] Ceccarelli C and Dominik C 2006 Astrophys. J. **640** L131

Cosmological constraints from supernova data set with corrected redshift

A. Feoli[1], **L. Mancini**[2,3,4], **V. Rillo**[1], **M. Grasso**[1]

[1] Department of Engineering, University of Sannio, Piazza Roma 21, 82100 – Benevento, Italy
[2] Max Planck Institute for Astronomy, Königstuhl 17, 69117 – Heidelberg, Germany
[3] Department of Physics, University of Salerno, Italy
[4] International Institute for Advanced Scientific Studies, Vietri sul Mare (SA), Italy

E-mail: feoli@unisannio.it

Abstract. Observations of distant type Ia supernovae (SNe Ia), used as standard candles, support the notion that the Cosmos is filled with a mysterious form of energy, the *dark energy*. The constraints on cosmological parameters derived from data of SNe Ia and the measurements of the cosmic microwave background anisotropies indicate that the dark energy amounts to $\approx$ 70% of all the energy contained in the Universe. In the hypothesis of a flat Universe ($\Omega_{\rm m} + \Omega_\Lambda = 1$), we investigate if the dark energy is really required in order to explain the SNe Ia experimental data, and, in this case, how much of such unknown energy is actually deduced from the analysis of these data and must be introduced in the ΛCDM model of cosmology. In particular we are interested in verifying if the Einstein–de Sitter model of the expanding Universe is really to be ruled out. By using a fitting procedure based on the Newton method search for a minimum, we reanalyzed the "Union compilation" reported by Kowalski et al. (2008) formed by 307 SNe, obtaining a very different estimate of the dark energy, that is $\approx$ 60%. Furthermore, in order to balance the correction of the apparent magnitude of SNe Ia, due to the dilation or stretching of the corresponding light curve width, we introduce a suitable modified redsfhit. Taking into account this correction, we refitted the Union compilation dataset after a selection cut. The main result that emerges from our analysis is that the values of $\Omega_{\rm m}$ and Ω_Λ strongly depend on the fitting procedure and the selected sample. In particular, the constraint we obtain on the mass density, normalized by the critical mass density, is $\Omega_{\rm m} = 0.7$ for a sample of 252, and $\Omega_{\rm m} = 1$ for a sample of 242 SNe Ia respectively. The latter case does not imply the existence of any additional form of dark energy.

1. Introduction

It is well known that the supernovae (SNe) Ia have a significant intrinsic brightness due to their very homogenous initial conditions, i.e. thermonuclear explosions of white dwarfs. For this reason they are among the best standard candles providing cosmological distance determination, with a precision of approximately 8%, and thus determination of cosmological parameters [1]. The detection of SNe Ia thus turns out to be essential in order to probe the cosmological expansion (for a review, see ref. [2]).

After the first systematic efforts to detect low–redshift ($z < 0.1$) SNe Ia, carried out by the Calàn/Tololo Supernova Survey [3, 4, 5], several international teams, the Supernova Cosmological Project (SCP) and the High-Z SN Search (HZSNS) performed extensive searches for SNe Ia at high redshift (out to $z \leq 0.8$) [6, 7, 8, 9, 10]. In the wake of these successful

Published under licence by IOP Publishing Ltd

experiments, a fit of the magnitude-redshift data of 42 SNe Ia (at redshifts between 0.18 and 0.83), discovered by the Supernova Cosmology Project (SCP), jointly with a set of 18 supernovae from the Calàn/Tololo Supernova Survey, was performed at redshifts below 0.1 [11]. Practically, this study compared the distance modulus of the measured SNe Ia with that which would be expected for an Einstein–de Sitter flat Universe, leading to the truly surprising discovery of a positive vacuum energy density ($\Omega_\Lambda > 0$), i.e. the *dark energy*, which opposes the self–attraction of matter and causes the expansion of the Universe to accelerate (for a review, see refs. [12, 13]). This result is consistent with the full sky temperature maps based on seven years of data from WMAP, which are well fit by a minimal six-parameter flat ΛCDM model. In fact, by using the WMAP data in conjunction with baryon acoustic oscillation data from the Sloan Digital Sky Survey and priors on the Hubble constant H_0 from Hubble Space Telescope observations, it was possible to estimate that $\Omega_\Lambda = 0.728^{+0.015}_{-0.016}$ [14].

In more recent years, several large-scale ground-based projects have been started [15, 16, 17, 18, 19], the Hubble Space Telescope playing a key role in high-precision optical and infrared follow-up of SNe discovered from the ground. It was therefore possible to draw up new SN Ia compilations, combining high-redshift ($z \sim 0.5$) with low-redshift ($z \sim 0.05$) samples, collected in different projects [20, 21, 22, 23, 24, 25, 26]. However, the corresponding cosmological constraints were consistent with a very wide range of dark energy models. This fact is basically caused by statistical errors but, most importantly, by systematic uncertainties as well. As a matter of fact, the statistical uncertainties in measurements of dark energy parameters are on the same order as systematic uncertainties. Moreover, the different results significantly depend on which light curve fitter has been used. This fact clearly emerges from subsequent works [27, 28, 29, 30]. It is also remarkable to note that the constraints on the value of Ω_Λ, which is the main goal of this research, depend strongly on redshift of the SNe Ia. In fact, the data at $z > 1$ are not able to provide robust constraints on the existence of the dark energy. Lastly, the physical nature of the dark energy is at the moment lost in the meanders of theoretical speculations [31, 32, 33, 34, 35, 36, 37, 38].

In the present paper, we challenge the current standard model of cosmology, the ΛCDM model, which includes the cosmological constant as an essential feature. In particular we debate on the use of the stretch factor that intervenes in the SN Ia light curves to modify the time axis around the peak of the luminosity. Actually, without this empirical transformation, the light curves of different SNe Ia are not congruent with each other, but have a shape and a maximum that differ substantially among the samples. However, we believe that any correction applied on the apparent magnitude of each SNe Ia implies a correction on its corresponding redshift, as we discuss in section 2. From this perspective we have reanalysed the so–called "Union compilation", which is a compilation of 414 SNe Ia (which reduces to 307 after selection cuts) provided by ref. [27] together with a framework to analyze datasets in a homogeneous manner[1].

The paper is structured as follows. In §2 we introduce the correction that we carried on the redshift of SNe. In §3 we explain the fitting procedure and the selection cuts on the original sample that have led to the compilation of two subsets. The resulting parameters of the fits are reported together with the Hubble diagrams. Finally, in §4 we summarize and discuss the results.

2. Apparent magnitudes and Redshift corrections

The apparent magnitudes of SNe measured by different groups at different redshifts must be corrected, before putting them together in the Hubble diagram, in such way that the absolute magnitude can be considered the same constant for all the SNe forming the sample. The photometric images are corrected for galactic extinction, K–correction, etc., and finally undergo

[1] Recently this analysis chain has been further refined in ref. [39], the "Union2 sample".

a light-curve width-luminosity correction [6]. The light curve of each SN is compared with a template and the peak apparent magnitude is corrected assuming a functional form [11]

$$m^{\text{eff}} = m^{\text{peak}} + \alpha(s-1), \tag{1}$$

where α is a coefficient that will be determined by a fitting algorithm, simultaneously with the other cosmological parameters; s is the so called "stretch factor" that suitably stretches or contracts the time axis of each light curve around the date of maximum light, affecting both the rising and declining part of the light curve of each SN.

The stretch factor was introduced to reduce the intrinsic dispersion in the absolute brightness of supernovae, but it really acts as a physical mechanism and it was first considered due to a local astrophysical effect (for example the temperature - dependent variation in the opacity) by the authors in ref. [6]. Then, four years later, they same authors admitted that: "*a single stretch factor varying the timescale of the SNe Ia accounts very well for the restframe B band light curve both before and after maximum light*" but "*It is not understood from the current status of the theory of SNe Ia whether this is a fortuitous coincidence or a reflection of some physics timescale*" [40].

The behavior of the SNe Ia light curves is usually interpreted as a strong support for time dilation due to an expanding Universe. However, non-standard cosmological models have been suggested as alternative explanations for acceleration without dark energy. As a matter of fact, if we use a model-independent method to compare quantitatively the ability of some dark energy models and non-standard types of cosmology (braneworld, $f(R)$, kinematic models, etc.) to represent the Union2 sample data [39], none of the considered cosmological models can be rigorously rejected on the basis of the current SN Ia data [41].

Here we propose an alternative view of the stretch factor that we interpret as a virtual displacement of the supernova from the measured redshift to a new effective redshift. Actually the width of the light curves is larger for SNe Ia at higher redshift ($z > 0.1$) than it is for local ones. In fact, due to redshift, the observed width evolves by a *time-dilatator* factor $(1+z)$. Still the same authors [40] adjusted the SCP and Calàn/Tololo photometric data to the maximum intensity I_{max} and scaled the time axis by a width factor $w = s(1+z)$. Remarkably, with the application of this single stretch timescale parameter, the SNe Ia data fell all on a common curve at the level of the measurement uncertainty.

However, in our opinion, the correction of the apparent magnitude should be balanced by a correspondent correction of the redshift. In fact, since the transformation from the observer frame to the SN rest frame is obtained by dividing the time interval by the factor $w = s(1+z)$, this is equivalent to consider a time-dilatator factor $(1+z_{\text{corr}}) = s(1+z)$, and SNe shall not be considered at redshift z anymore, but at somewhat corrected redshift z_{corr}, such that

$$z_{\text{corr}} = s + sz - 1. \tag{2}$$

In the light of our interpretation of the stretch factor, we think that the SN Ia data have to be refitted by using the above corrected redshift, which takes into account nothing else than the correction of the apparent magnitude of SNe adducted by the scientists of the SCP on the photometric data.

3. Hubble diagram construction and cosmological parameter fitting

We considered the data reported in ref. [27], even if these authors did not strictly preserved the original meaning of s. Actually, in the recent update of the SNe sample [39] the meaning of the stretch factor was completely lost because the new fitting algorithm SALT2 returns a parameter x_1 that must be converted to the old s using a correlation formula that would become a source of other statistical errors.

The effective magnitude of the i-th of the N SNe forming our modified sample is obtained from the formula:

$$m_i^{\rm eff} = m_i^{\rm peak} + \alpha(s_i - 1) - \beta c_i \tag{3}$$

where c is the rest frame color at maximum as defined in ref. [27] and β is an additional parameter, which has been determined by the fitting procedure too.

The χ^2 corresponding to that of eq.(3) is given as

$$\chi^2 = \sum_{i=1}^{N} \frac{(m_i^{\rm eff} - 5\log_{10} D_i - \widetilde{M})^2}{(\sigma_{m\,i} + \alpha\,\sigma_{s\,i} - \beta\,\sigma_{c\,i})^2 + \sigma_{\rm int}^2 + \sigma_{v\,i}^2}, \tag{4}$$

where D is the Hubble free luminosity distance

$$D = (1+z)\int_0^z \frac{dz'}{\sqrt{\Omega_m(1+z')^3 + \Omega_\Lambda}}, \tag{5}$$

$\widetilde{M}$ is the magnitude zero point offset

$$\widetilde{M} = M + 5\log_{10}\frac{C}{H_0} + 25, \tag{6}$$

C is the speed of light, and M is the absolute magnitude which is assumed to be constant after that the above mentioned corrections of apparent magnitudes have been implemented. σ_m, σ_s, σ_c are the errors related to the magnitude, stretch factor, and color factor, respectively. We neglected the errors on z and $z_{\rm corr}$. As in ref. [42], we considered an error due to a peculiar velocity dispersion of 300 km sec^{-1}, which can be written:

$$\sigma_{v\,i} = \frac{0.005}{\ln 10}\left(\frac{1}{1+z_i} + \frac{1}{\sqrt{\Omega_{\rm m}(1+z_i)^3 + \Omega_\Lambda}\int_0^{z_i}\frac{dz}{\sqrt{\Omega_{\rm m}(1+z)^3+\Omega_\Lambda}}}\right). \tag{7}$$

Furthermore, we supposed that our Universe is flat, so that $\Omega_{\rm m} + \Omega_\Lambda = 1$.

We found a minimum of our χ^2 in eq.(4) starting from an intrinsic dispersion of the relation $\sigma_{\rm int} = 0.15$, and we updated this value refitting until we have obtained a reduced $\chi^2_{\rm red} = \chi^2/{\rm DOF} \simeq 1$. To this aim we used the "Newton method search for a minimum" developed in Mathematica in a way very similar to the algorithm adopted in ref. [42]. The same results can be equivalently obtained using iteratively the "NonlinearRegress" functionality of Mathematica.

We applied this procedure to the standard sample of 307 SNe Ia reported in ref. [27]. The resulting four parameters ($\Omega_{\rm m}$, $\widetilde{M}$, α, β) of the fitting procedure are listed in Table 1, which clearly shows, in the first row, a difference of more than 10% in the estimate of $\Omega_{\rm m}$ between our procedure and the well known standard result.

Next, we corrected the measured redshift z by eq.(2), and after this correction, we neglected the 35 following SNe because their $z_{\rm corr}$ are now negative: 1993ag, 1993o, 1993h, 1992br, 1992bp, 1992bo, 1992bl, 1992aq, 1990af, 2001cn, 2000bh, 1993ac, 1994m, 1994t, 1995ak, 1996bo, 2000fa, 2000dk, 2000cn, 2000cf, 1999gd, 1999ek, 1999cc, 1998eg, 1998ef, 1998dx, 1998co, 1998ab, 1998v, 1997dg, 1997y, 1999bm, 1995ap, h283, n326.

Then, we considered the remaining 272 SNe, and we made a fitting test without taking into account the measurement errors, and imposing $\Omega_{\rm m} = 1$. Actually, we did not want a really blind analysis of the data, but we wanted to test if the Einstein-de Sitter model, without any cosmological constant, should really be excluded from the cosmological theory. Considering the

Journal of Physics: Conference Series **354** (2012) 012005 doi:10.1088/1742-6596/354/1/012005

Table 1. Fitting procedure results.

z	N	$\Omega_{\rm m}$	$\widetilde{M}$	α	β	$\sigma_{\rm int}$	$\chi^2_{\rm red}$
normal	307	0.393 ± 0.046	24.015 ± 0.035	0.81 ± 0.12	1.15 ± 0.10	0.25	1.056
corrected	252	0.70 ± 0.22	24.24 ± 0.11	11.01 ± 0.41	1.25 ± 0.28	0.32	1.005
corrected	242	1	24.400 ± 0.034	10.57 ± 0.33	1.23 ± 0.24	0.20	1.075

parameters resulting from the test fit, 20 SNe were removed from the 272 SN sample having the highest residuals with respect to the Einstein-de Sitter model; they are: 1992ag, 1992ae, 1992p, 1990o, 1999gp, 1994s, 1999aa, 1999ao, 1999aw, 1996j, 1996cl, 2001jn, 2001jf, 040mb, 04Sas, 2003dy, e138, g142, k396, p534. After this selection cut, we applied our fitting procedure considering also the measurement errors and the corresponding results are listed in the second line of Table 1. With respect to the 307 SN data set, we obtained a completely different estimation of the mass density, i.e. $\Omega_{\rm m} = 0.7$.

In order to verify the stability of our result, we removed other 10 SNe from our sample (1992bc, 2001cz, 1995ac, 1995ao, 2002w, 04D3df, 03D1cm, bo10, g055, m138). This time, the value of $\Omega_{\rm m}$ changed again and became equal to the unity, causing the vanishing of any form of dark energy. Note that the minimum of χ^2 in the last case falls outside the range of the allowed values $0 \leq \Omega_{\rm m} \leq 1$. So we have computed some three parameters fits for fixed values of $\Omega_{\rm m}$ in the right range. The result is that χ^2 is a decreasing function of $\Omega_{\rm m}$ that has the smallest acceptable value just when we put $\Omega_{\rm m} = 1$. The values of the remaining three parameters are reported in Table 1. The behavior of the Hubble diagrams are shown in figures 1 and 2 for the 252 and the 242 SN subset, respectively.

By inspection of Table 1, it is evident that the absolute magnitude and the parameter β are stable for the three samples. Instead, the cosmological parameter $\Omega_{\rm m}$ changes drastically, and also the parameter α derived using $z_{\rm corr}$ is very different from the one obtained with the standard z. The reason of the latter difference can be explained expanding "$D(z_{\rm corr})$" in the equation (5) around $s = 1$. At the first order in $(s-1)$ we have

$$5 \log_{10} D(z_{\rm corr}, \Omega_{\rm m}) \simeq 5 \log_{10} D(z, \Omega_{\rm m}) + \delta(z, \Omega_{\rm m})(s-1), \tag{8}$$

where

$$\delta = \frac{5}{\ln 10}\left(1 + \frac{1+z}{\sqrt{\Omega_{\rm m}(1+z)^3 + \Omega_\Lambda}\int_0^z \frac{dz'}{\sqrt{\Omega_{\rm m}(1+z')^3 + \Omega_\Lambda}}}\right). \tag{9}$$

If we insert this result in the equation (4) we can write $\chi^2(z_{\rm corr})$ in terms of the normal z. Comparing it with the corresponding standard $\chi^2(z)$ we obtain $\alpha_{\rm norm} = \alpha_{\rm corr} - \delta(z_i)$ in the sum. So, considering $\bar{\delta} = \sum_{i=1}^N \delta(z_i)/N$, which is the average value of δ in our sample, the estimates of α listed in Table 1, when we use the corrected redshift, are in effect $\alpha_{\rm corr} \simeq \alpha_{\rm norm} + \bar{\delta}$.

4. Conclusions

We introduced a correction on the redsfhit of the SNe Ia that takes into account the correction of the corresponding luminosity, which is commonly performed in order to gets all the SN light curves to match. In this framework, we reanalyzed the Union compilation [27] finding the following results:

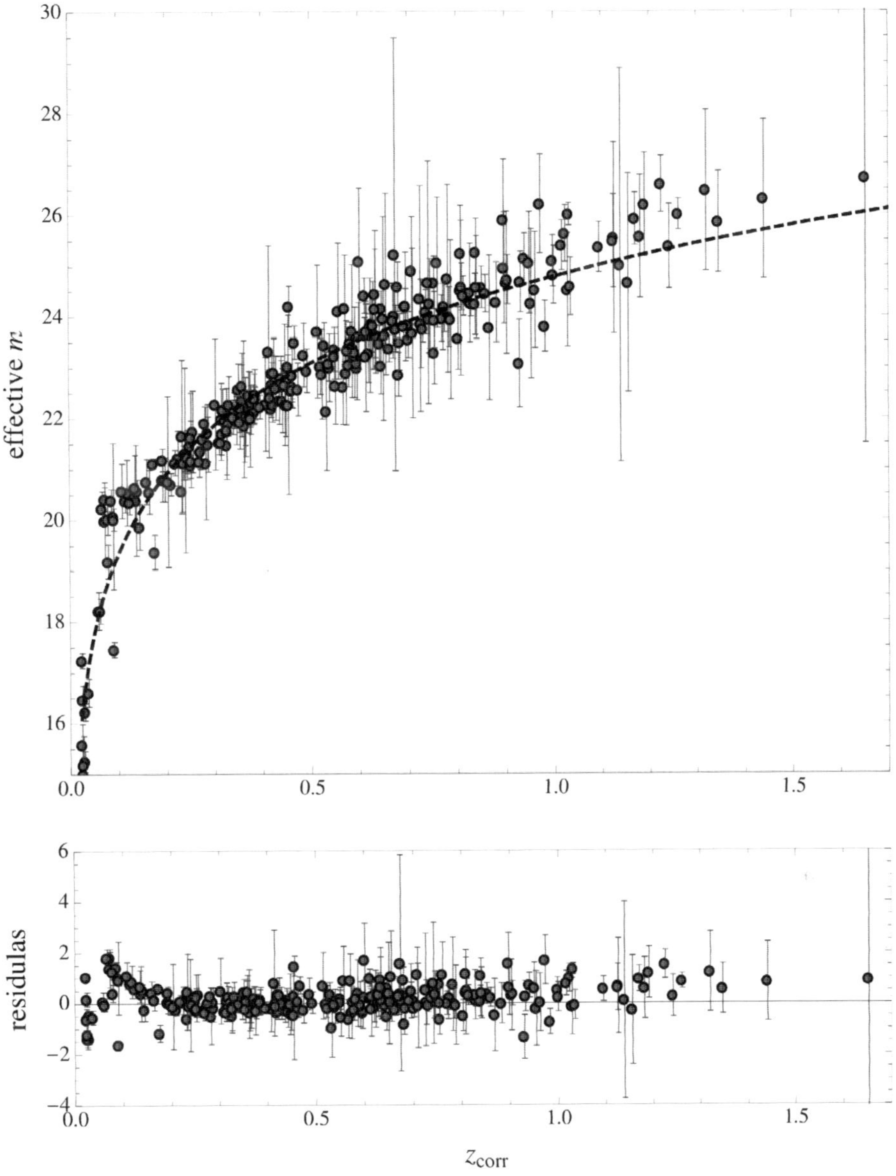

Figure 1. Top: Hubble diagram for 252 type Ia SNe selected (see text) from the "Union compilation" reported in [27]. Bottom: Magnitude residuals from the best-fit flat cosmology, ($\Omega_m = 0.70$, $\Omega_\Lambda = 0.30$).

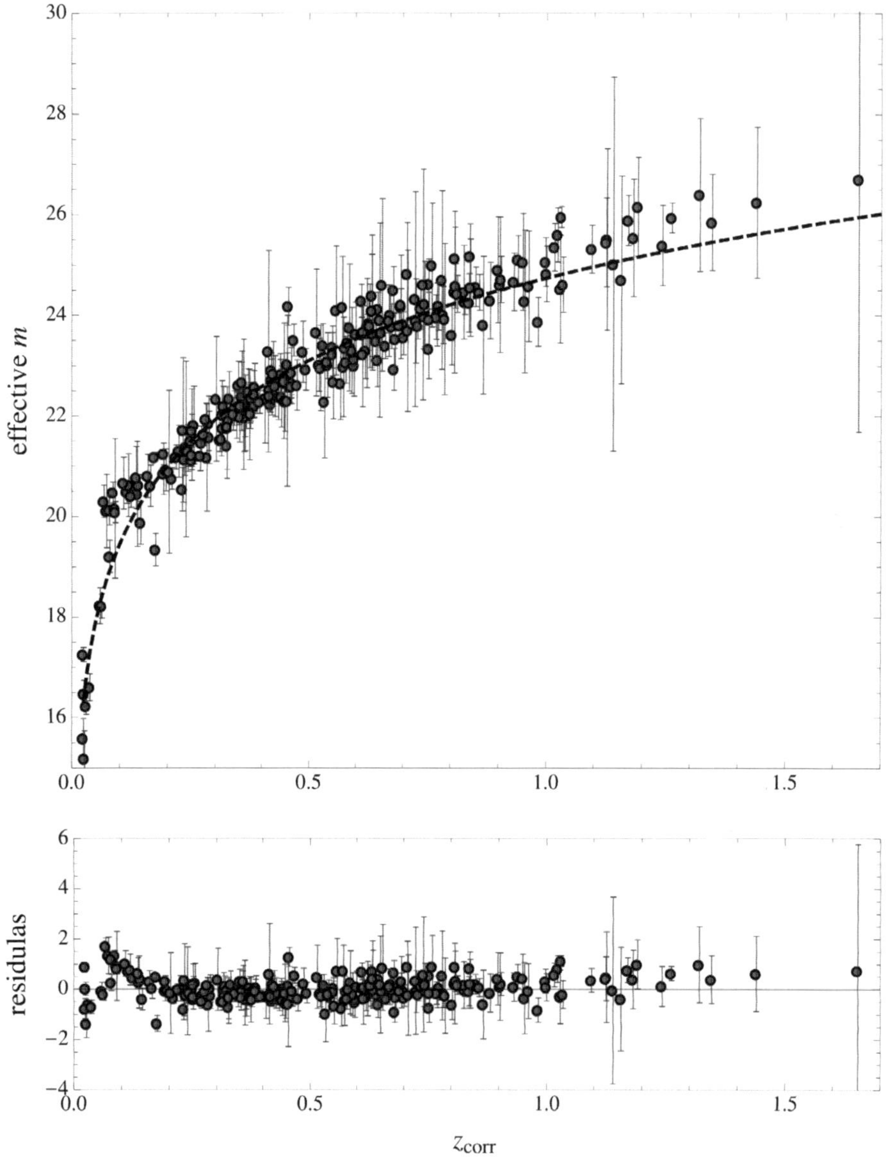

Figure 2. Top: Hubble diagram for 242 type Ia SNe selected (see text) from the "Union compilation" reported in [27]. Bottom: Magnitude residuals from the best-fit flat cosmology, ($\Omega_m = 1$, $\Omega_\Lambda = 0$).

- We refitted the set of 307 type Ia SNe by a fitting procedure based on the Newton method search for a minimum. The constraint we obtained from supernovae on the dark energy density is $\Omega_\Lambda = 0.607 \pm 0.046$, which is quite different from the result of $\Omega_\Lambda = 0.713 \pm 0.028$ found in ref. [27]. This outcome reflects the fact that the fit of the SNe is very sensible to the fitting procedure and to the systematic and intrinsic errors inserted in the χ^2.
- We corrected the redshift of the SNe according to eq. (2), and neglected the SNe whose redshift turned out to be negative. After a selection cut based on the residuals, we reduced the original compilation to two subsets formed by 252 and 242 SNe respectively. Fitting the first subset, we found that the values of the cosmological parameters were practically reversed ($\Omega_{\rm m} = 0.70$, $\Omega_\Lambda = 0.30$). On the other hand, from the fit of the second subset we find even $\Omega_{\rm m} = 1$, which corresponds to a Universe without dark energy.
- The fits of the SNe, both with z and $z_{\rm corr}$, depend on the number of SNe used too. The difference of few SNe between two samples changes dramatically the estimates of the cosmological parameters. The above mentioned case of the two subsets formed by 252 and 242 SNe respectively is emblematic.

In conclusion, we found that the constraint on the cosmological parameters based on the analysis of SNe is contingent on both fitting procedure and the number of SNe considered in the sample. On the contrary, the absolute magnitudes of SNe are not affected by variations in the sample.

We emphasize that it is well known that the hypothesis according to which the Universe is accelerating comes not only from the results of SNe Ia, but also from the analysis of CMB and galaxy clusters. The latter two methods converge on a result of $\Omega_\Lambda \simeq 0.70$, $\Omega_{\rm m} \simeq 0.30$. The present work is not intended to replace the current paradigm of the cosmology (even if the debate is still alive [43]), but only to point out that the use of SNe Ia in the estimate of cosmological parameters can introduce, through the stretch factor, some instability in the final result, in addition to biases and discrepancies in the SN Ia data samples. So, all the results that come from the SNe Ia must be taken with more caution, especially when they are used to discriminate between different cosmological models.

Acknowledgments

The data of the original *Union compilation* has been downloaded from http://supernova.lbl.gov /Union. Our numerical analysis has benefited from the Mathematica files of S. Nesseris and L. Perivolaropoulos, which are available on http://leandros.physics.uoi.gr/snls.htm. A.F. and L.M. acknowledge support for this work by research funds of the University of Sannio.

References

[1] Perlmutter S 2003 Supernovae, Dark Energy, and the Accelerating Universe *Physics Today* **56** 53
[2] Perlmutter S and Schmidt B P 2003 Measuring Cosmology with Supernovae *Lecture Notes in Physics* **598** 195
[3] Hamuy M et al 1993 The 1990 Calan/Tololo Supernova Search *Astron. J.* **106** 2392
[4] Hamuy M et al 1995 A Hubble diagram of distant type IA supernovae *Astron. J.* **109** 1
[5] Hamuy M et al 1996 BVRI Light Curves for 29 Type IA Supernovae *Astron. J.* **112** 2408
[6] Perlmutter S et al 1997 Measurements of the Cosmological Parameters Omega and Lambda from the First Seven Supernovae at $Z \geq 0.35$ *Astrophys. J.* **483** 565
[7] Perlmutter S et al 1998 Discovery of a supernova explosion at half the age of the universe *Nature* **391** 51 [Erratum 1998 **392** 311]
[8] Garnavich P M et al 1998 Constraints on Cosmological Models from Hubble Space Telescope Observations of High-z Supernovae, *Astrophys. J. Letter* **493** L53
[9] Schmidt B P et al 1998 The High-Z Supernova Search: Measuring Cosmic Deceleration and Global Curvature of the Universe Using Type IA Supernovae *Astrophys. J.* **507** 46

[10] Riess A G et al Observational Evidence from Supernovae for an Accelerating Universe and a Cosmological Constant *Astron. J.* **116** 1009
[11] Perlmutter S et al 1999 Measurements of Omega and Lambda from 42 High-Redshift Supernovae *Astrophys. J.* **517** 565
[12] Frieman J, Turner M and Huterer 2008 Dark Energy and the Accelerating Universe *Annu. Rev. Astron. Astrophys.* **46** 385
[13] Blanchard A 2010 Evidence for the fifth element. Astrophysical status of dark energy *Astron. Astroph. Rev.* **18** 595
[14] Jarosik N et al 2011 Seven-year Wilkinson Microwave Anisotropy Probe (WMAP) Observations: Sky Maps, Systematic Errors, and Basic Results *Astrophys. J. Suppl.* **192** 14
[15] Filippenko A V, Li W D, Treffers R R, Modjaz M, The Lick 2001 Observatory Supernova Search with the Katzman Automatic Imaging Telescope *Proc. Astron. Soc. Pac. Conference Series* **246**, *IAU Colloq.* **183**: *Small Telescope Astronomy on Global Scales*, ed. Paczynski B, Chen W-P and Lemme C, p 121
[16] Nobili S et al 2005 Restframe I-band Hubble diagram for type Ia supernovae up to redshift $z \sim 0.5$ *Astron. Astroph.* **437** 789
[17] Copin Y et al 2006 The Nearby Supernova Factory *New Astron. Rev.* **50** 436
[18] Hamuy M et al 2006 The Carnegie Supernova Project: The Low-Redshift Survey *Publ. Astron. Soc. Pac.* **118** 2
[19] Law N M et al 2009 The Palomar Transient Factory: System Overview, Performance, and First Results, *Publ. Astron. Soc. Pac.* **121** 1395
[20] Tonry J L et al 2003 Cosmological Results from High-z Supernovae *Astrophys. J.* **594** 1
[21] Knop R A et al 2003 New Constraints on Ω_M, Ω_Λ, and w from an Independent Set of 11 High-Redshift Supernovae Observed with the Hubble Space Telescope *Astrophys. J.* **598** 102
[22] Barris B J et al 2004 Twenty-Three High-Redshift Supernovae from the Institute for Astronomy Deep Survey: Doubling the Supernova Sample at $z > 0.7$ *Astrophys. J.* **602** 571
[23] Riess A G et al, Type Ia Supernova Discoveries at $z > 1$ from the Hubble Space Telescope: Evidence for Past Deceleration and Constraints on Dark Energy Evolution *Astrophys. J.* **607** 665
[24] Astier P et al 2006, The Supernova Legacy Survey: measurement of Ω_M, Ω_Λ and w from the first year data set *Astron. Astroph.* **447** 31
[25] Wood-Vasey W M et al 2007 Scrutinizing Exotic Cosmological Models Using ESSENCE Supernova Data Combined with Other Cosmological Probes *Astrophys. J.* **666** 694
[26] Riess A G et al 2007 New Hubble Space Telescope Discoveries of Type Ia Supernovae at $z \geq 1$: Narrowing Constraints on the Early Behavior of Dark Energy *Astrophys. J.* **659** 98
[27] Kowalski M et al. 2008 Improved Cosmological Constraints from New, Old, and Combined Supernova Data Sets *Astrophys. J.* **686** 749
[28] Holtzman J A et al 2008 The Sloan Digital Sky Survey-II: Photometry and Supernova IA Light Curves from the 2005 Data *Astron. J.* **136** 2306
[29] Hicken M et al 2009 Improved Dark Energy Constraints from 100 New CfA Supernova Type Ia Light Curves *Astrophys. J.* **700** 1097
[30] Kessler R et al. 2009 First-Year Sloan Digital Sky Survey-II Supernova Results: Hubble Diagram and Cosmological Parameters *Astrophys. J. Suppl.* **185** 32
[31] Copeland E J, Sami M and Tsujikawa S 2006 Dynamics of Dark Energy *Int. J. Mod. Phys. D* **15** 1753
[32] Ellis G 2008 *Cosmology: Patchy solutions Nature* **452** 158
[33] Buchert T 2008 *Dark Energy from structure: a status report, Gen. Rel. Grav.* **40** 467
[34] Durrer R and Maartens R 2008 Dark energy and dark gravity: theory overview *Gen. Rel. Grav.* **40** 301
[35] Linder E V 2008 Resource Letter DEAU-1: Dark energy and the accelerating universe *American J. of Physics* **76** 197
[36] Linder E V 2008 The dynamics of quintessence, the quintessence of dynamics *Gen. Rel. Grav.* **40** 329
[37] Mattsson T 2010 Dark energy as a mirage *Gen. Rel. Grav.* **42** 567
[38] Kamionkowski M 2011 Dark matter and dark energy *Visions of Discovery: New Light on Physics, Cosmology, and Consciousness*, ed. by Chiao R Y, Cohen M L, Leggett A J, Phillips W D, Harper C L (Cambridge Univ. Press) p 247
[39] Amanullah R et al 2010 Spectra and Hubble Space Telescope Light Curves of Six Type Ia Supernovae at $0.511 < z < 1.12$ and the Union2 Compilation *Astrophys. J.* **716** 712
[40] Goldhaber G et al 2011 Timescale Stretch Parameterization of Type Ia Supernova B-Band Light Curves *Astrophys. J.* **558** 359
[41] Benitez-Herrera S. et al 2012 Model-independent reconstruction of the expansion history of the Universe from Type Ia supernovae *MNRAS* **419** 513
[42] Nesseris S and Perivolaropoulos L 2005 Comparison of the legacy and gold type Ia supernovae dataset constraints on dark energy models *Phys. Rev. D* **72** 123519
[43] Sarkar S 2008 Is the evidence for dark energy secure? *Gen. Rel. Grav.* **40** 269

Journal of Physics: Conference Series **354** (2012) 012006 doi:10.1088/1742-6596/354/1/012006

About microlensing optical depth and rates for free-floating planets towards the Kepler's field of view

M Hafizi[1] **and L Hamolli**
Physics Department, Faculty of Natural Sciences, Tirana University, Bul. Zogu I, Tirana, Albania

E-mail: mimoza.hafizi@fshn.edu.al

Abstract. In this work we examine the possibility of observing microlensing events in the Kepler space observatory field of view, caused by brown dwarfs or free-floating planets. We calculate the optical depth towards the field of view of the Kepler satellite and the rate of these events based on latest results about mass distribution of astrophysical objects from brown dwarf down to Earth mass order. With the current data, the probability of such events is insignificant, due to the small number of stars observed by this instrument compared to other experiments devoted to the microlensing method. Nevertheless, this probability may increase significantly in the case of a higher presence of free-floating planets, whose number is poorly defined so far.

1. Gravitational microlensing

The gravitational microlensing effect happens when the gravitational field of an astrophysical object acts as a lens to magnify background source stars ([1]). A gravitational lens of mass M is characterized by its Einstein ring radius, the radius of the ring image created when the observer, the lens at distance D_l and the source at distance D_s are perfectly aligned, which is defined as:

$$R_E(M,x)=\sqrt{\frac{4GM}{c^2}D_s x(1-x)}\,, \tag{1}$$

where $x=\dfrac{D_l}{D_s}$ is the normalized lens distance. The image separation for astrophysical bodies in galactic distances is too small to be resolved. In such microlensing events, the observable feature is the light magnification due to the lens-source relative motion with transverse velocity v_t. The microlensing technique has already shown its great potentiality and has allowed the detection of thousands of events towards galactic bulge, several tens of events towards the LMC, SMC and M31 galaxy and 11 exoplanets until now.

The key parameter of a microlensing light curve is the Einstein radius crossing time given by:

[1] To whom any correspondence should be addressed.

Published under licence by IOP Publishing Ltd

$$t_E = \frac{R_E}{v_t}, \tag{2}$$

which can be considered as the characteristic event duration. Its value towards the Galactic centre is of the order of $t_E \propto \sqrt{\frac{M}{M_J}}$ days, $M_J = 9.5\times10^{-4} M_S$ being the Jupiter mass, M_S the solar mass. Microlensing events with $t_E < 2$ days indicate, so, planetary-mass lenses. Thus, although t_E depends on the velocity and distance of the lens object, in addition to its mass, the t_E distribution can be used to probe statistically the mass function of the lens objects.

The microlensing optical depth is defined as the probability that at any time a random star is magnified more than a threshold amplification $A_{th} = 1.34$ by a certain population of lenses (i.e., thin and thick stellar disk, white and brown dwarfs or planetary mass populations). It is given by:

$$\tau_i = \frac{4\pi G D_s^2}{c^2} \int_0^1 \rho_i(x) x(1-x) dx, \tag{3}$$

where $\rho_i(x) = m_i n_i(x)$ is the lens mass density of the i-th lens object galactic population ([2]).

The microlensing rate Γ_i, i.e. the number of events per unit time and per monitored star due to i-th lens object population is given by:

$$\Gamma_i = 2\int R_E(x) n_i(x) v_L f(\vec{v}_L - \vec{v}_t) f(\vec{v}_s) dx\, d\vec{v}_L d\vec{v}_s, \tag{4}$$

where $\vec{v}_L, \vec{v}_s, \vec{v}_t$ are the lens, source and microlensing tube two-velocities in the plane transverse to the line of sight ([2]). The velocity distribution functions $f(\vec{v}_L - \vec{v}_t)$ and $f(\vec{v}_s)$ are assumed to have a Maxwellian form with one-dimensional dispersion velocity σ_i different for each lens and source population.

2. Free-floating planetary populations

In a recent work ([3]) is reported the discovery of a population of unbound Jupiter-mass objects, known as free-floating planets and whose existence has been questioned several times during this decade ([4,5]). After two years of gravitational microlensing surveys towards Galactic Bulge performed by MOA (Microlensing Experiments in Astrophysics) and OGLE (Optical Gravitational Lensing Experiment) groups, 10 microlensing events with $t_E < 2$ days are obtained out of 474 with t_E from fractions of a day to more than a hundred days. The lenses for these 10 short duration events are inferred to be either unbound planets or bound with wide separations of more than ten astronomical units from their host stars, so that one cannot detect the host star in the light curves. For this reason, they are called free-floating planets.

By considering a broken power-law mass function $\frac{dN}{dM} = M^{-\alpha_i}$ with $\alpha_1 = 2$ for $0.7 < M/M_S < 1$, $\alpha_2 = 1.3$ for $0.08 < M/M_S < 0.7$ (the initial stars with masses above the solar mass are assumed to have evolved into stellar remnants) and α_3 as a fitting parameter for the brown dwarf regime $0.01 < M/M_S < 0.08$, the ten microlensing events with $t_E < 2$ lead the authors of [3] to a fourth population, that is a planetary population with masses $M/M_S < 0.01$ having a mass function $\frac{dN}{dM} = M^{-\alpha_{PL}}$, with $\alpha_{PL} = 1.3^{+0.3}_{-0.4}$, much steeper than the brown dwarf slope $\alpha_3 = 0.49^{+0.24}_{-0.27}$. The abrupt change in the mass function favours the idea that the formation process of this planetary population is different from that of stars and brown dwarfs. They may have formed in proto-planetary disks and subsequently scattered into unbound or very distant orbits.

The broken power-law mass function analysis indicates that the number of brown dwarfs would be $0.73^{+0.22}_{-0.19}$ times the number $N_* = 2\times10^{11}$ of the main sequence stars in the mass range $0.08 < M / M_S < 1$. The model also predicted a larger, but poorly constrained number of planetary mass objects per star with $N_{PL} / N_* = 5.5^{+18.1}_{-4.3}$, due to weak sensitivity of observations to very low-mass lenses. The lower mass limit for planets is taken $10^{-5} M_S$.

A consequence of the above mentioned work is a kind of correlation between variations of α_3 and α_{PL}, which we parameterize in our analysis by an approximate linear relation:

$$\alpha_{PL} = -\alpha_3 + 1.7. \qquad (5)$$

3. Free-floating planets towards Kepler's field of view

In this work we are looking for the possibility to see microlensing events produced by brown dwarfs or free-floating planets and detected by Kepler mission. Although the principal goal of this mission is the detection of bound planets transiting their host stars, its capacity to analyze fine light fluctuations might lead to identification of eventual microlensing events. The Kepler Space observatory is placed at a heliocentric orbit, it is designed to survey main sequence stars at a portion of sky of 115 square degrees, with galactic coordinates centered at $b = 13.5°$, $l = 76.32°$ and lying from 150 to 1000 pc. The total number of stars in its field of view is about half million, 150 000 of them observed simultaneously every 30 minutes.

In this analysis we calculate the microlensing optical depth and the rate of microlensing events towards this field of view for two populations, brown dwarfs (BD) and free-floating planets (PL), based on relations (3) and (4) corrected with the mass distribution function $f(M)$ for this case of lenses with a certain range of masses.

The data we need are spatial and velocity distributions for lenses and sources. For the two lens populations we assume the thin disk spatial distribution (see [6] for more details):

$$\rho(M, R, z) = f(M) e^{-|z|/H} e^{-(R-R_0)/h}, \qquad (6)$$

with $H = 300pc$, $h = 3.5kpc$, $R_0 = 8.5kpc$ and $f(M) = \frac{dN}{dM} = M^{-\alpha_i}$ described above. For the velocity distribution we consider $\sigma_{PL} = \sigma_{BD} = 50km/s$ ([7]).

Concerning the sources, the observed stars lie in a depth of the order of 1 kpc, so we take into account their double exponential thin and thick luminous disk distribution:

$$\rho_d(R, z) = \frac{\Sigma_0}{2H} e^{-|z|/H} e^{-(R-R_0)/h}, \qquad (7)$$

with $\Sigma_0 = 25M_S pc^{-2}$, $H = 0.3kpc$, $h = 3.5kpc$ for the thin component and $\Sigma_0 = 35M_S pc^{-2}$, $H = 1kpc$, $h = 3.5kpc$ for the thick component.

Instead of the fixed lens mass, we make use of the broken power-law mass function $f(M) = M^{-\alpha_i}$ with respectively $\alpha_3 = 0.49^{+0.24}_{-0.27}$ and $\alpha_{PL} = 1.3^{+0.3}_{-0.4}$, as concluded in [3]. The lower value of the mass range is taken of the order of $10^{-5} M_S$, corresponding to the lens mass which could produce a microlensing event with a characteristic duration 30 minutes, the lower value of the observation interval for the Kepler instrument. The higher limit of the free-floating mass range gets to the brown dwarf lower limit $0.01M_S$.

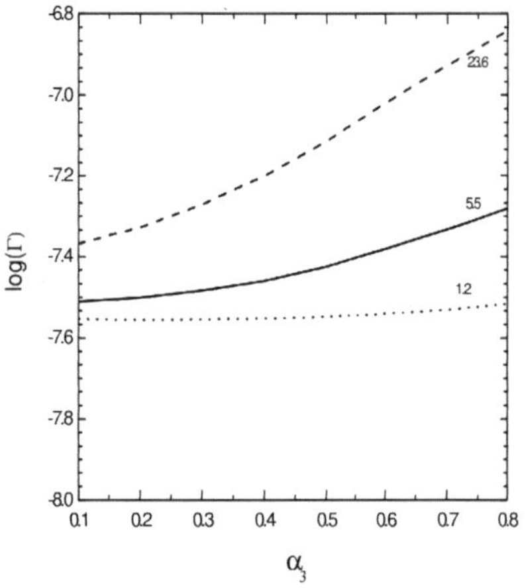

Figure 1. The microlensing rate expected for brown dwarf and free-floating planetary disk models shown as a function of brown dwarfs mass function index α_3. The dotted line corresponds to the minimum of the planetary number $N_{PL} / N_* = 1.2$, the continuous to the most plausible value $N_{PL} / N_* = 5.5$ and the dashed line to the maximum value of the planetary number $N_{PL} / N_* = 23.6$.

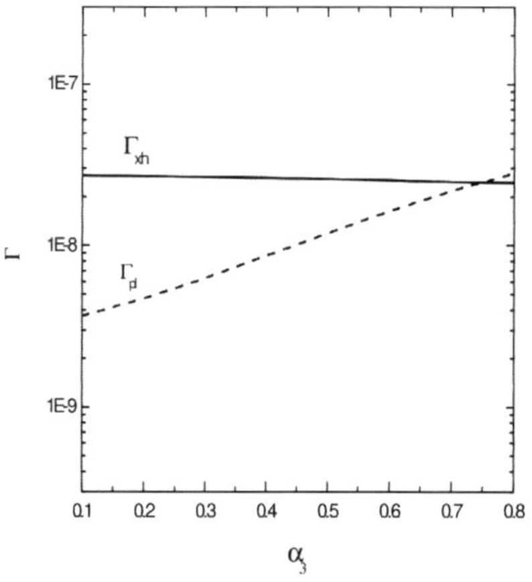

Figure 2. Here we show the importance of the planetary population in microlensing rate (dashed line) compared to that of the brown dwarf population (continuous line). We obtain that the planetary population becomes dominant in such events for higher values of α_3.

4. Model results

In figure 1 and 2 we present the results we obtained for the microlensing rate dependence on brown dwarf mass function index, this last one varying in the interval of values found in [3]: $0.1 < \alpha_3 < 0.8$. Owing to relation (5), free-floating planet mass function index varies also in the range $0.9 < \alpha_{PL} < 1.6$. For the brown dwarf number we keep a fixed value in the process of mass function calibration, the value $0.73N_*$ mentioned above. On the contrary, the free-floating planet number varies in a wide range, $N_{PL} / N_* = 5.5^{+18.1}_{-4.3}$. We need thus to reflect the results for extreme limits of this number. We show in figure 1 three cases of free-floating planet abundance: two extreme values $N_{PL} = 1.2N_*$, $N_{PL} = 23.6N_*$ and the most probable one $N_{PL} = 5.5N_*$.

In figure 1 we present the microlensing rate expected for both brown dwarf and free-floating planetary disk model. We obtain rate values increasing with the mass index, therefore we conclude that steeper mass functions for brown dwarfs (and softer for planets) give higher probabilities to get microlensing event signs in Kepler observations.

In figure 2 we show the weight of free-floating planets in the microlensing rate in comparison to that of brown dwarfs. Here we have chosen the most plausible value for the planetary abundance, $N_{PL} = 5.5N_*$. We find that this rate increases with the mass index, whereas the rate for brown dwarfs

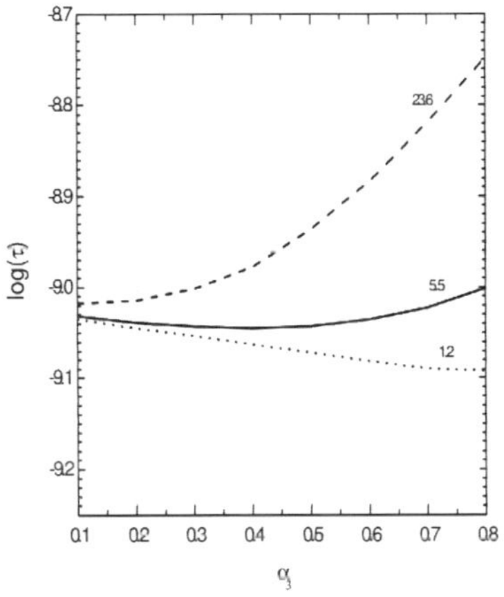

Figure 3. The optical depth for brown dwarf and free floating planetary disk model shown as a function of brown dwarfs mass function index α_3. The dotted line corresponds to the minimum of the planetary number $N_{PL}/N_* = 1.2$, the continuous to the most plausible value $N_{PL}/N_* = 5.5$ and the dashed one to the maximum value of the planetary number $N_{PL}/N_* = 23.6$.

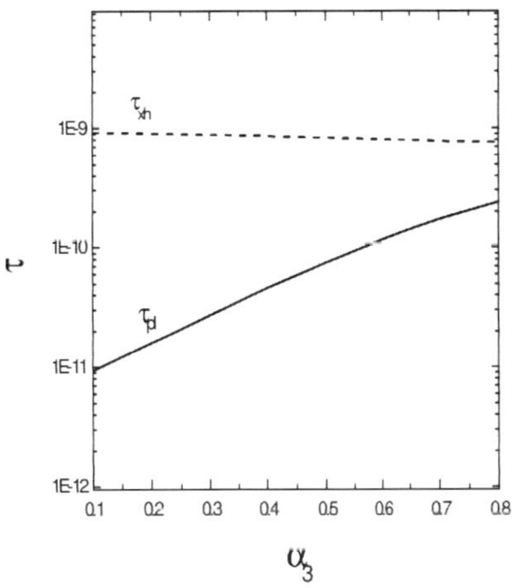

Figure 4. Here we show the importance of the planetary population in optical depth (continuous line) compared to that of the brown dwarfs population (dashed line). We obtain that the planetary population is at least one order less important.

decreases very softly. For values of the index above 0.7, the eventual microlensing events would be caused mainly by free-floating planet lenses.
In figure 3 and 4 we repeat for the optical depth the same calculations as above. The gap in the optical depth between the two considered populations is higher than in microlensing rate.

Once the microlensing rate for each lens population is calculated, it is straightforward to obtain the expected number of microlensing events by the simple multiplication of the rate with the number of observed stars. The most optimistic value we find is about 0.022 events per year, attained in the case of the upper limit for the number of free-floating planets.

5. Conclusions

With 150 000 stars observed by Kepler mission, it is obvious to obtain very small probabilities of detecting microlensing events during its time of observation. However, the present analysis is relevant since the detection of even a short-time microlensing event (for example by the forthcoming space-based telescopes like WFIRST and/or EUCLID) would require to review our work-hypotheses and look for:

1. A steeper mass function for brown dwarfs, associated to a softer one for free-floating planets, which leads to higher values of microlensing rate and optical depth.
2. A higher presence of free-floating planets than the limit found up to now.
3. Different models for their spatial and velocity distributions.

References

[1] Paczynski B 1986 *Astrophys. J.* **304** 1
[2] Griest K 1991 *Astrophys. J.* **366** 412
[3] Sumi T 2011 *Nature* **473/7347** 349
[4] Smith K W and Bonnello A 2001 *Mon.Not. Roy. Astron. Soc.* **322** L1
[5] Zinnecker H 2001 *Preprint arXiv: astro-ph/0107351v1*
[6] Hafizi M, de Paolis F and Nucita A 2004 *Int. J. Mod. Phys.* D **13/9** 1831
[7] Jetzer Ph, Mancini L and Scarpetta G 2002 *Astron. Astrophys.* **393** 129

Centre-of-Mass Energy of Colliding Particles in the Background of Charged Black Strings

IBRAR HUSSAIN

School of Electrical Engineering and Computer Science
National University of Sciences and Technology
H-12, Islamabad, Pakistan

E-mail: ibrar.hussain@seecs.nust.edu.pk

Abstract. The centre-of-mass (CM) energy of collision for two particles falling freely from rest at infinity is investigated in the background of a charged black string. It is found that like black holes, the CM energy of collision is arbitrarily high in the vicinity of the horizon of the black string if one of the colliding particles has critical charge and the other particle takes any different value of the charge.

Key words: Charged black strings; Particles collision; Centre-of-mass energy

1. Introduction

Bañados, Silk and West (BSW) found that infinite centre-of-mass (CM) energies can be produced during the particles collision in the vicinity of the event horizon of the extremal Kerr black hole [1]. A kinematic explanation of acceleration of particles by black holes can be found in [2]. A general explanation for the effect of unbound acceleration of particles by black holes is avaliable in [3]. After the seminal work of BSW, their mechanism was then applied to a series of black hole spacetimes [4, 5, 6, 7, 8, 9, 10, 11, 12, 13, 14, 15, 16, 17, 18, 19, 20, 21, 22, 23]. Berti et al. and Jacobson and Sotiriou obtained some practical limitations on the achievable CM energy from the astrophysical limitations [24, 25].

Following the same sprite, in the next section we study the CM energy of collision for two charged particles falling freely from rest at infinity in the background of a charged black string. We show that like black holes, the charged black string can also accelerate the colliding particles to an unlimited CM energy if one of the colliding particles has critical charge and the other particle takes any different value of the charge.

We conclude our discussion in the last section. Here we use the system of units $G = c = 1$.

2. Charged Black Strings and Collision Energy in the CM Frame

The Hoop conjecture, proposed by Kip Thorne in 1972 [26], states that horizons form when and only when mass M of an object gets compressed into a region whose circumference in every direction is less than its Schwarzschild circumference, $4\pi M$. This excluded the possibility of the formation of cylindrical black holes. However, the Hoop conjecture holds only if the cosmological constant vanishes. Thus in the presence of a negative cosmological constant cylindrical black holes (also called black strings) [27], have been investigated with great interest. They represent asymptotically anti-de Sitter (AdS) spaces in the transverse and axial directions. Here we discuss the collision process for the charged black string. The line element of the charged black string is given by [28]

$$ds^2 = -fdt^2 + \frac{1}{f}dr^2 + r^2 d\phi^2 + \alpha^2 r^2 dz^2, \tag{1}$$

where

$$f = \alpha^2 r^2 - \frac{4M}{\alpha r} + \frac{4Q^2}{\alpha^2 r^2}, \tag{2}$$

and $\alpha^2 = -\Lambda/3 > 0$, denotes the negative cosmological constant. In (2) M and Q are the Arnowitt-Deser-Misner (ADM) mass and charge per unit length in the z direction. The event horizon can be obtained by putting $f = 0$, as [29]

$$r_\pm = \frac{(4M)^{\frac{1}{3}}}{2\alpha}\Big[\sqrt{s} \pm \sqrt{2\sqrt{s^2 - Q^2(\frac{2}{M})^{\frac{4}{3}}} - s}\Big]. \tag{3}$$

Here $r_\pm$ represents the outer and inner horizons and the quantity s in (3) is defined by

$$s = \Big(\frac{1}{2} + \frac{1}{2}\sqrt{1 - \frac{64Q^6}{27M^4}}\Big)^{\frac{1}{3}} + \Big(\frac{1}{2} - \frac{1}{2}\sqrt{1 - \frac{64Q^6}{27M^4}}\Big)^{\frac{1}{3}}. \tag{4}$$

The singularity at $r = 0$ is surrounded by the horizons $r_\pm$ if the inequality $Q^2 \leqslant \frac{3}{4}M^{\frac{4}{3}}$, holds. The black string will be extremal if

$$Q^2 = \frac{3}{4}M^{\frac{4}{3}}. \tag{5}$$

The electric potential is defined by

$$A = \frac{2Q}{\alpha r^2}. \tag{6}$$

Consider two charged particles having the same rest mass m are at rest at infinity and then they start to move toward the charged black sting defined above. Let these particles collied at some distance r. Let q_1 and q_2 be the charge of the two incoming particles respectively. The collision energy in the CM frame is defined by

$$E_{cm} = m\sqrt{2}\sqrt{1 - g_{\mu\nu}u_1^\mu u_2^\nu}, \tag{7}$$

where

$$u^\mu = \frac{dx^\sigma}{d\tau}(\frac{\partial}{\partial x^\sigma})^\mu \tag{8}$$

is the four-velocity of each of the colliding particle at the point of collision [1]. The motion of particles can be determined by the following Lagrangian [30]

$$L = \frac{1}{2}g_{\mu\nu}\dot{x^\mu}\dot{x^\nu} + \frac{qA_\mu}{m}\dot{x}^\mu, \tag{9}$$

where A_μ is the electromagnetic 4-potential and $\dot{x^\mu} = dx^\mu/d\tau$.

For the charged black string (1) the Lagrangian (9) becomes

$$L = \frac{1}{2}\Big[-f\dot{t}^2 + \frac{1}{f}\dot{r}^2 + r^2\dot{\phi}^2 + \alpha^2 r^2\dot{z}^2\Big] - \frac{2Qq}{m\alpha r^2}\dot{t}. \tag{10}$$

This Lagrangian does not depends explicitly on t, therefore the following conserved quantity exists

$$\frac{\partial L}{\partial \dot{t}} = -f\dot{t} - \frac{2Qq}{m\alpha r^2} = -\frac{E}{m}, \tag{11}$$

where E is the energy of the particle as measured by an observer at rest at infinity. We take for simplicity $\dot{\phi} = 0 = \dot{z}$.

From (11) we write

$$\dot{t} = \frac{1}{f}\Big(\frac{E}{m} - \frac{2Qq}{m\alpha r^2}\Big). \tag{12}$$

The normalization condition

$$g_{\mu\nu}u^\mu u^\nu = -1, \tag{13}$$

yields

$$\dot{r} = -\sqrt{\Big(E - \frac{2Qq}{\alpha r^2}\Big)^2\frac{1}{m^2} - f}. \tag{14}$$

The substitution of (12) and (14) in (7) gives

$$E_{cm} = \sqrt{2}m\Big[1 + \frac{1}{f}\Big(\frac{E_1}{m} - \frac{2Qq_1}{m\alpha r^2}\Big)\Big(\frac{E_2}{m} - \frac{2Qq_2}{m\alpha r^2}\Big)$$
$$-\frac{1}{f}\sqrt{\Big(\frac{E_1}{m} - \frac{2Qq_1}{m\alpha r^2}\Big)^2 - f}\sqrt{\Big(\frac{E_2}{m} - \frac{2Qq_2}{m\alpha r^2}\Big)^2 - f}\Big]^{\frac{1}{2}} \tag{15}$$

This CM energy expressed by (15) blows up at the horizon i.e. at $f = 0$. At $f = 0$ the numerator in (15) also vanishes. Therefore, we calculate the limiting value of the E_{cm} at the event horizon by expanding (15) around $f = 0$ i.e. around r_+, as follows

$$E_{cm} = \sqrt{2}m\sqrt{1 + \frac{1}{2}\Big(\frac{q_2 - \frac{E_2 r_+^2 \alpha}{2Q}}{q_1 - \frac{E_1 r_+^2 \alpha}{2Q}} + \frac{q_1 - \frac{E_1 r_+^2 \alpha}{2Q}}{q_2 - \frac{E_2 r_+^2 \alpha}{2Q}}\Big)}. \tag{16}$$

The expression (16) clearly show that the CM energy which depends on the charge of the black string will be unlimited in the vicinity of the horizon if one of the colliding particles has the critical charge

$$q_c = \frac{E\alpha r_+^2}{2Q}, \tag{17}$$

and the other colliding particle takes any other value of the charge. This result is similar to the results obtained for the Reissner-Nordstrom and Reissner-Nordstrom de-Sitter black holes [4, 16]. For the colliding particles to reach the horizon of the black string the quantity inside the square root in (14) must be positive.

3. Conclusion

Here we have investigated the CM energy of collision for two particles falling freely from rest at infinity in the background of a charged black string. We have found that the CM energy of collision of two charged particles is arbitrarily high at the event horizon of the black string. This arbitrarily high value of the CM energy of collision can be obtained at the horizon if one of the colliding particles has the critical value of charge given by (17) and the other particle takes any different value of the charge. Here we see that, in order to accelerate particles to an arbitrarily high CM energy of collision the black string need not be an extremal one.

Acknowledgments

This work has been supported by Higher Education Commission of Pakistan (HEC) and French Embassy joint split Ph.D. and Post-Doctoral fellowship programme. I would like to thank Laboratoire Jacques-Louis Lions of the Universitè Pierre et Marie Curie (Paris 6), Paris, where the writing-up of this paper was completed. I would also like to thank the unknown referee for giving useful comments on this work.

References

[1] M. Banados, J. Silk, S.M West, Phys. Rev. Lett. **103**, 111102 (2009).

[2] O. B. Zaslavskii, Phys. Rev. **D 84**, 024007 (2011).

[3] O. B. Zaslavskii, Class. Quantum. Grav. **28**, 105010 (2011).

[4] O. B. Zaslavskii, JETP Lett. **92**, 571 (2010).

[5] A. A. Grib and Y. V. Pavlov, JETP Lett. **92**, 125 (2010).

[6] S. W. Wei, Y. X. Liu, H. Guo and C. E Fu, Phys. Rev. **D 82**, 103005 (2010).

[7] K. Lake, Phys. Rev. Lett. **104** 211102 (2010).

[8] O. B. Zaslavskii, Phys. Rev. **D 81**, 044020 (2010).

[9] S. W. Wei, Y.X. Liu, H. Guo, Chun E Fu, Phys. Rev. **D 82**, 103005 (2010).

[10] S.W. Wei, Y.X. Liu, H. Guo, Chun-E Fu, JHEP 1012 066 (2010).

[11] Y. Zhu, S. Wu, Y. Jiang, G. Yang, arXiv:1108.1843.

[12] A. Williams, Phys. Rev. **D 83**, 123004 (2011).

[13] M. Patil, A. S. Joshi, K. Nakao, M. Kimura, arXiv: 1108.0288.

[14] S. Gao and C. Zhong, Phys. Rev. **D 84**, 044006 (2011).

[15] P. J. Mao, R. Li, L. Y. Jia and J. R. Ren, arXiv:1008.2660v3.

[16] C. Zhong and S. Gao, JETP Lett. **94**, 631 (2011).

[17] Y. Li, J. Yang, Y. Li, S. Wei, Y. Liu, arXiv:1012.0748v2.

[18] T. Harada and M. Kimura, Phys.Rev. **D 83**, 024002 (2011).

[19] T. Harada and M. Kimura, Phys.Rev. **D 83**, 084041 (2011).

[20] Y. Zhu, S. Wu, Y. Liu and Y. Jiang, Phys. Rev. **D 84**, 043006 (2011).

[21] M. Patil and P. S. Joshi, Phys. Rev. **D 84**, 104001 (2011).

[22] M. Patil and P. S. Joshi, Class. Quantum Grav. **28**, 235012 (2011).

[23] I. Hussain, Mod. Phys. Lett. A (to appear).

[24] E.Berti, V.Cardoso, L.Gualtieri, F.Pretorius, and U.Sperhake, Phys. Rev. Lett. **103**, 239001 (2009).

[25] T. Jacobson, T. P.Sotiriou, Phys. Rev. Lett. **104**, 021101 (2010).

[26] K. Thorne, Nonspherical Gravitational Collapse: a Short Review in Magic without Magic, ed. J. Klauder (San Francisco: Freeman) (1972).

[27] J. P. S. Lemos, Phys. Rev. **D 54**, 3840 (1996).

[28] R. G. Cai and Y. Z. Zhang, Phys. Rev. **D 54**, 4891 (1996).

[29] A. Fatima and K. Saifullah, arXiv:1108.1622v1.

[30] D. Pugliese, H. Quevedo, R. Ruffini, Phys. Rev. **D 83**, 104052 (2011).

Fractional Action Cosmology with Power Law Weight Function

Mubasher Jamil
Center for Advanced Mathematics and Physics, National University of Sciences and Technology, H-12, Islamabad, Pakistan
E-mail: mjamil@camp.nust.edu.pk

Muneer A. Rashid
Center for Advanced Mathematics and Physics, National University of Sciences and Technology, H-12, Islamabad, Pakistan
E-mail: muneerrshd@yahoo.com

D. Momeni
Department of Physics, Faculty of Sciences, Tarbiat Moa'llem University, Tehran, Iran
E-mail: d.momeni@yahoo.com

Olga Razina
Eurasian International Center for Theoretical Physics, Eurasian National University, Astana 010008, Kazakhstan

Kuralay Esmakhanova
Eurasian International Center for Theoretical Physics, Eurasian National University, Astana 010008, Kazakhstan

Abstract. Motivated by an earlier work on fractional-action cosmology with a periodic weight function [1], we extend it by choosing a power-law weight function in the action. In this approach, we obtain a varying gravitational coupling constant. We then model dark energy in this paradigm and obtain relevant cosmological parameters.

1. Dirac's Large Numbers Hypothesis and time running of G

During our survey of various physical laws of nature, we came across a number of constants entangled with these laws. Historically it was Weyl [2, 3] who initiated the idea of large numbers. However, it was Dirac who discovered an apparently unseen thread joining up these physical constants by a simple yet interesting law, viz., *Law of Large Numbers*. Using this law, Dirac arrived at his *Large Number Hypothesis* (LNH) which had profound influence on the world of physics. It generated a lot of activity both at theoretical and observational level with LNH as their starting point. Some works are centered around modification of Einstein's gravitational theory and related equations to adapt the G variation [4, 5]. Some authors have also given counter-arguments against LNH for justifying and refuting that hypothesis characterize the second category of works in favor of it [6] and also testing the validity of it [7]. Even there is found a formal link between LNH and cosmological constant term Λ [8]. Weyl's large number is square root of Eddington's number $N \sim 10^{80}$ (which represents the total number of protons in the Universe) [9] and is supported by observational data [10]. Indeed Stewart showed that the ratio of the radius of the Universe and the electron is only two orders of magnitude smaller 10^{40} than that of Weyls number. There are some other large numbers like Jordan's number [11] and the Shemi-zadeh's number [12] which have also appeared in the literature.

As originally stated by Dirac there are three dimensionless numbers in nature which can be constructed from the atomic and cosmological data:

(i) The ratio of the electric to the gravitational force between an electron and a proton $\frac{e^2}{Gm_e m_p} \simeq 7 \times 10^{39}$.

(ii) The mass of that part of the Universe that is receding from us with a velocity $v < c/2$ expressed in units of the proton mass is 10^{78}.

(iii) The age of the Universe t, expressed in terms of a unit of time $\frac{e^2}{m_e c^3}$ provided by atomic constants, is inversely proportional to Newton's gravitational constant.

The above three statements imply in respective order

$$\frac{e^2}{Gm_e m_p} \propto t, \tag{1}$$

$$M \propto t^2, \tag{2}$$

$$G \propto t^{-1}. \tag{3}$$

There is a serious problem between (3) and GR [4, 13]: Einstein's theory requires G to be a constant. As was noted by Dirac, this inconsistency might be resolved if "*.. we assume that Einstein theory is valid in a different system of units from those provided by the atomic constants.*" Consequence of Dirac's LNH is the existence of a variable G cosmology. A beautiful account of the constants of nature appears in [14].

2. Framework

Recently Nabulsi [15] examined the Friedmann-Robertson-Walker (FRW) spacetime in the context of fractional action-like variational approach or fractionally differentiated Lagrangian function depending on a fractality parameter χ to model nonconservative and weak dissipative classical and quantum dynamical systems.

We apply the technical machinery of differential geometry to general relativity in the context of fractional action-like variational approach. Following Einstein's arguments, we start by representing the gravitational field by an affine connection on a curved manifold and assume that the free fall corresponds to geodesic motion. Thus the classical Lagrangian of a curve is defined by $L = (1/2)g_{ij}\dot{x}^i\dot{x}^j$, where $\dot{x}^i = dx^i/dt$ and g_{ij} is the metric tensor. This gives the action

$$S = \frac{m}{2}\int_0^\tau g_{ij}(x)\dot{x}^i\dot{x}^j\mu(\chi, t)dt, \tag{4}$$

where m is the particle's mass. Here $\mu(\chi, t)$ is the generalized weight function and if we choose it to be a periodic function, we recover the results in [1]. This weight function can produce a friction force in classical dynamics for the test particle and as we will show, it also yields a time varying G. The modified resultant fractional Euler-Lagrange equations are

$$\ddot{x}^l + \partial_t \log(\mu(\chi, t))\dot{x}^l + \Gamma^l_{ij}\dot{x}^i\dot{x}^j = 0. \tag{5}$$

We follow the procedure of [15] and assume that the test particle's velocity is approximately zero. Furthermore, the metric and its derivative are approximately static. Applying these assumptions to the spatial components of (5) gives:

$$\partial_i \overrightarrow{\gamma}^i + \partial_t \mu \partial_i v^i + R_{tt} = 0. \tag{6}$$

In the above equation $v^i = dx^i/dt$ and $\gamma^i = dv^i/dt$ are the particle's velocity and acceleration respectively and $\partial_i\gamma^i = -4\pi G\rho$, which is Poisson's equation in Newtonian gravity. Here ρ is the mass density in gravitational interaction and G is Newton's gravitational constant. The resultant modified gravitational constant is

$$G' = G\Big(1 - \frac{\partial_t \log(\mu(\chi, t))}{4\pi G\rho}\partial_i v^i\Big) = G + \triangle G. \tag{7}$$

If $\mu(\chi, t) = t$ then (7) describes Dirac's model [13]. Thus the change in G in a spatially flat $(D + 2)$-dimensional FRW Universe is

$$\triangle G = -\frac{D(D+1)H\partial_t \log(\mu(\chi, t))}{8\pi\rho},$$

where $H = \partial_t(\log(a(t))$ is the Hubble parameter. The Friedmann equation for such a model with metric $g_{ij} = diag(1, -a^2(t)\eta_{AB})$, $A, B = 2, ..., D + 2$, with $\eta_{AB} = diag(1, ..., 1)$ is

$$\frac{D(D+1)}{2}H^2 = 8\pi G\rho + \Lambda. \tag{8}$$

One thinks that if the gravitational constant changes slightly, as we have expressed it in this work, the Friedmann's equation must be replaced by an effective G. A simple example of effective G is the $f(R)$ gravity. An interesting feature of these theories is the fact that the gravitational constant is both time and scale dependent. After some calculations, one can obtain a Poisson equation in the Fourier space and attribute the extra terms as effective gravitational constant G_{eff} [16]. But now, we are working using an Einstein-Hilbert action therefore there is no need to insert effective G on the right hand side of the Friedmann's equation (8).

Using the Bianchi identity and assuming both energy density ρ and the cosmological constant Λ as functions of time related by $\Lambda(t) = \Lambda_0\rho(t)^n$. Consequently we have

$$G(t) = \frac{G}{\mu(\chi, t)}, \quad \Lambda(t) = \Lambda_n\mu(\chi, t)^{n/(1-n)}, \quad \rho(t) = \rho_n\mu(\chi, t)^{1/(1-n)}, \tag{9}$$

where $\Lambda_n \equiv \Lambda_0(\frac{8\pi G(1-n)}{\Lambda_0 n})^{n/(n-1)}$ and $\rho_n \equiv (\frac{8\pi G(1-n)}{\Lambda_0 n})^{1/(n-1)}$. Substituting (9) in (8) we obtain

$$H = \frac{\dot{a}}{a} = c_n\mu(\chi, t)^{n/(2(1-n))}, \tag{10}$$

where $c_n = \sqrt{\frac{2b_n}{D(D+1)}}$ and $b_n = 8\pi G\rho_n + \Lambda_n$ and $0 < n < 1$. By solving (10), we find

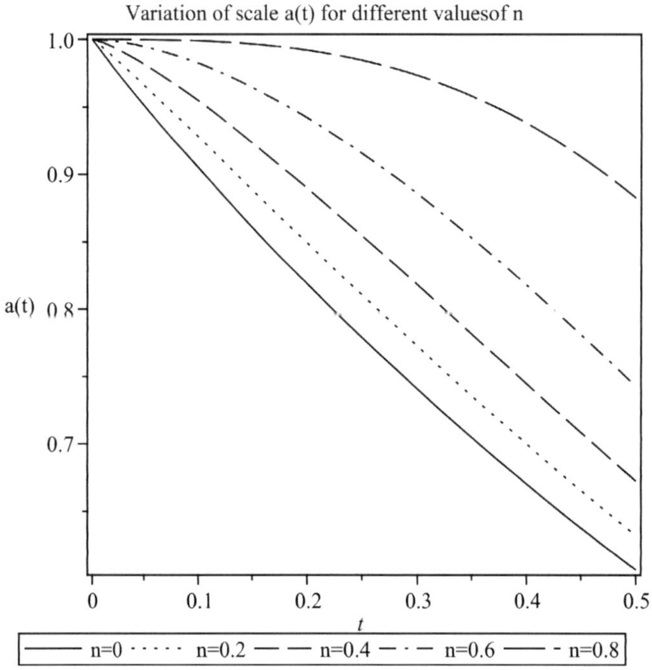

Figure 1. Variation of scale factor (13) for different values of n.

$$a(t) = a_0 \exp\Big(c_n \int \mu(\chi, t)^{n/(2(1-n))} dt\Big). \tag{11}$$

For the remaining part of the paper, we use power law weight function as

$$\mu(\chi, t) = \chi t^p, \qquad p \geq 1. \tag{12}$$

One can treat (12) as a generalized Dirac's hypothesis. Note that $p = 1$ leads to an inverse power of time relation for the gravitational constant i.e. $G = \frac{G_0}{t}$. Substituting (12) in (11) yields

$$a(t) = a_0 \exp(\alpha t^\beta), \tag{13}$$

where

$$\alpha \equiv \chi^{\frac{n}{2(1-n)}} \frac{2c_n(1-n)}{2+(p-2)n}, \quad \beta \equiv \frac{2+(p-2)n}{2(1-n)}. \tag{14}$$

It is useful to discuss the behavior of the scale factor diagrammatically. For simplicity we take $a_0 = 1$ and $\alpha = -1$, and focuss only on Dirac's model i.e. $G(t) = \frac{G}{t}$. Figure 1 shows the behavior of the scale factor as a function of time for different values of n.

The Hubble parameter is

$$H = \alpha\beta t^{\beta-1} = \alpha\beta\Big(-\frac{1}{\alpha}\ln(1+z)\Big)^{1-\frac{1}{\beta}}. \tag{15}$$

Here redshift z is defined by $1 + z = \frac{a_0}{a(t)}$. Figure 2 shows the behavior of the Hubble parameter as a function of time for $\alpha = 1$.

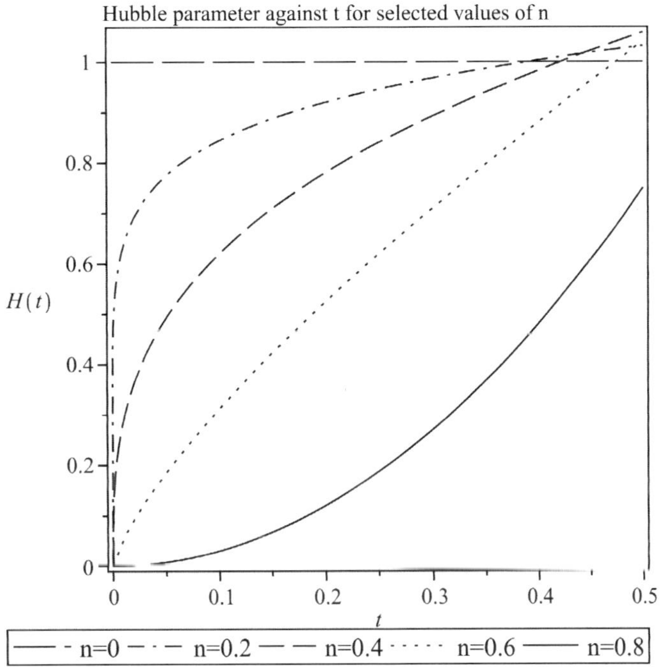

Figure 2. Hubble parameter against t for selected values of n.

We note that the scale factor for $n = \frac{2}{3}$, $\alpha = \frac{1}{2}$ is $a \propto \exp(\frac{t^2}{2})$. Consequently, the scale factor would increase very rapidly for the positive branch, while it would start from infinity for the negative branch. Also the choice $n = \frac{2}{3}$, $\alpha = \frac{-1}{2}$ gives $a \propto \exp(\frac{-t^2}{2})$. This indicates that the Universe is evolving with a Hubble parameter $H \sim -t$ and a scale factor $a \propto \exp(\frac{-t^2}{2})$, which is close to singularity related to the origin and the fate of the Universe. The same phenomena occurs in the context of a new model of dark energy, the Quintom model [17].

2.1. Deceleration parameter

We know that

$$q = -1 - \frac{\dot{H}}{H^2}, \tag{16}$$

is the deceleration parameter. For an expanding Universe, $q > -1$. Using (14) we can write (16) as

$$q = -1 - \frac{1-\beta}{\beta}\frac{1}{\ln(1+z)}. \tag{17}$$

For Dirac's hypothesis ($p = 1$), we have

$$q = -1 + \frac{n}{2-n}\frac{1}{\ln(1+z)}. \tag{18}$$

Thus we conclude that the Universe can remain in an accelerating expansion for some values of z. The next diagram (Figure 3) represents the graph of q versus z for different values of n. It

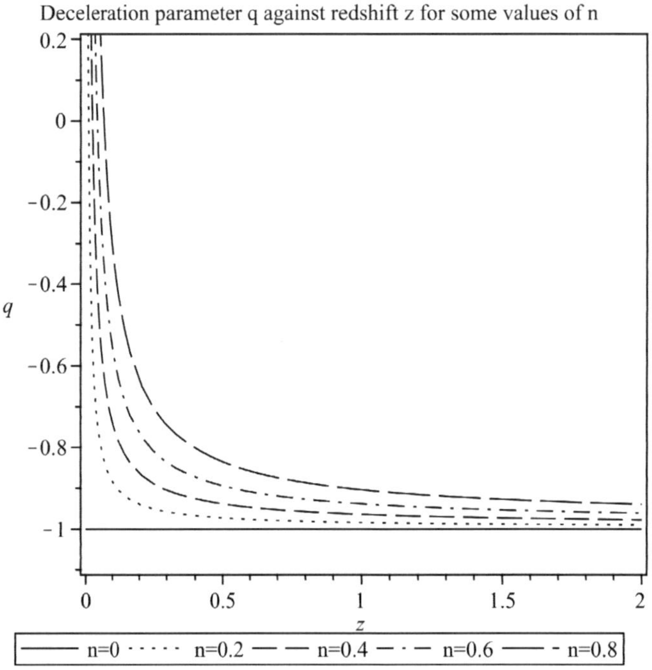

Figure 3. Deceleration parameter q against redshift z for some values of n.

shows that when n increases the deceleration parameter q recedes from the crossing line $q = -1$ [1]. Thus we can model cosmic acceleration in the present framework.

2.2. The statefinder parameters for dark energy

The statefinder parameters $\{r, s\}$ were introduced to characterize primarily the spatially flat Universe ($k = 0$) models with cold dark matter (dust) and dark energy [18]. They were defined as

$$r \equiv \frac{\dddot{a}}{aH^3} = \frac{(\beta^2 - 3\,\beta + 2)\,t^{-2\,\beta} + \beta\left((3\,\beta - 3)\,t^{-\beta} + \alpha\,\beta\right)\alpha}{\alpha^2\beta^2}, \tag{19}$$

$$s \equiv \frac{r-1}{3(q-\frac{1}{2})} = -\frac{2}{3}\,\frac{(\beta-1)\left(t^{-\beta}\beta - 2\,t^{-\beta} + 3\,\alpha\,\beta\right)}{\alpha\,\beta\,(3\,\alpha\,t^{\beta}\beta + 2\,\beta - 2)}. \tag{20}$$

From (19) and (20) we obtain the relation between r and s as

$$r = \frac{\alpha^{2-\beta}\,(2 + 7\,s\beta\,y - 4\,\beta - 9\,s + 16\,s\beta + 2\,s^2\beta^2 - 13\,s^2\beta - 7\,s\beta^2 - 7\,sy - 2\,\beta\,y + 2\,\beta^2 + 11\,s^2 + 2\,y)}{(-\beta + 1 - s\beta + s + y)^2}, \tag{21}$$

where

$$y = \sqrt{\beta^2 - 2\,\beta + 2\,s\beta + 1 - 2\,s + s^2\beta^2 - 2\,s^2\beta + s^2}. \tag{22}$$

[1] remembering that the $0 < n < 1$

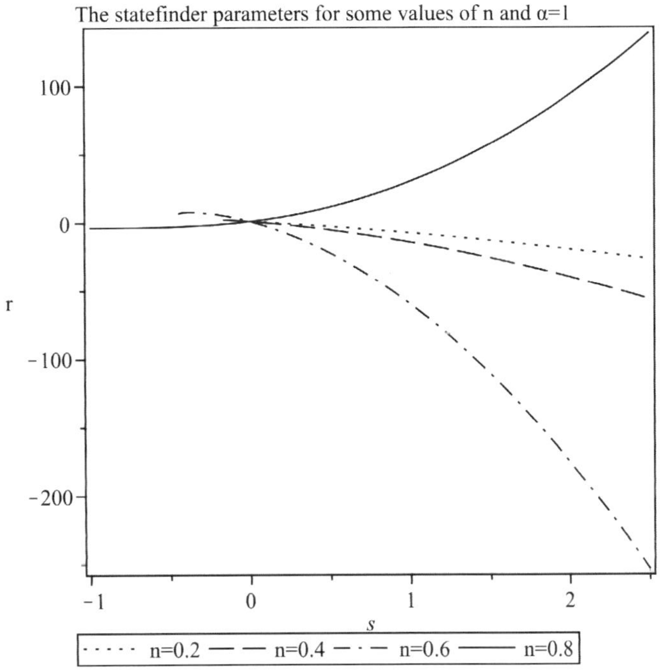

Figure 4. The statefinder parameters for some values of n.

We know that the cosmological constant possesses a fixed equation of state parameter ($w = -1$) and a fixed Newton's gravitational constant, hence $\{1,0\}$ corresponds to ΛCDM. Moreover $\{1,1\}$ represents the standard cold dark matter model containing no radiation while Einstein's static Universe corresponds to $\{-\infty,\infty\}$ [19].

Note that for Dirac's hypothesis $p = 1, n = 0, \beta = 1$, we have $\{r,s\} = \{1,0\}$ which represents a static cosmological constant with cold dark matter. In Figure-4, we draw the pair $\{r,s\}$ for different values of $\beta = \frac{2-n}{2(1-n)}, \alpha = 1$. As we can see, the fixed gravitational constant occurs several times for some values of n. Also our model describes the standard cold dark matter and a static Einstein Universe for some values of n.

2.3. Lookback time

If a photon is emitted by a source at the instant t and received at time t_0, then the photon travel time or the lookback time $t - t_0$ is defined by

$$t - t_0 = -\int_0^z \frac{dz}{H(z)(1+z)}, \tag{23}$$

where a_0 is the present value of the scale factor of the Universe. It is appropriate to calculate this time as a function of z. After substituting Eq. (15) in Eq. (23) we have

$$t - t_0 = -\frac{(-\alpha)^{1-1/\beta}}{\alpha\beta}\int_0^z \frac{dz}{(\log(1+z))^{1-1/\beta}(1+z)},$$

Now we change the variable as $x = \log(1+z)$. Calculating the integral above, we find

$$t - t_0 = (-\frac{\log(1+z)}{\alpha})^{\frac{1}{\beta}}. \tag{24}$$

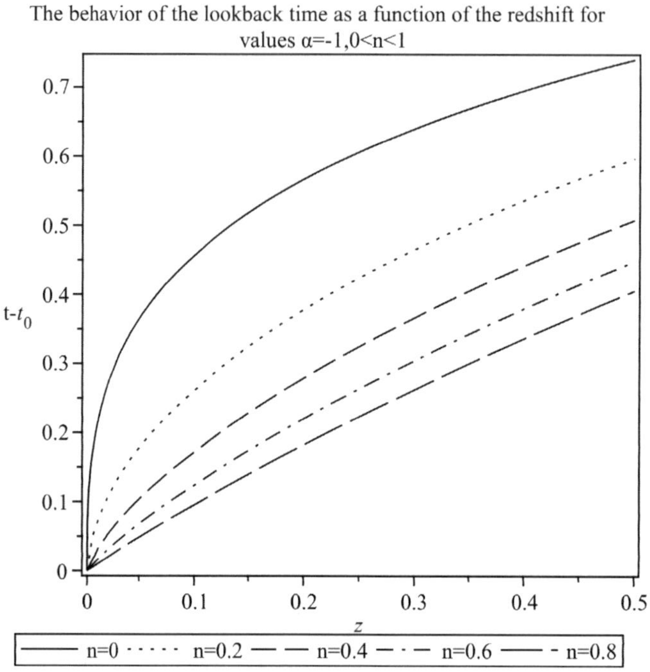

Figure 5. The behavior of the lookback time as a function of redshift for $\alpha = -1, 0 < n < 1$.

Figure-5 shows the behavior of the lookback time as a function of redshift for values $\alpha = -1, 0 < n < 1$.

2.4. *Proper distance*

If a photon is emitted by a source and is received by an observer at time t_0 then the proper distance between them is defined by

$$d = a_0 \int_a^{a_0} \frac{da}{aH^2} = \int_0^z \frac{dz}{H^2(z)(1+z)}. \tag{25}$$

Using Eq. (15) in (25), we get

$$d = \frac{1}{\beta(\beta-2)}(-\frac{\ln(1+z)}{\alpha})^{-1+\frac{2}{\beta}}. \tag{26}$$

Figure 6 shows the behavior of proper distance in the present model.

2.5. *Luminosity distance*

If L is the total energy emitted by the source per unit time and ℓ is the apparent luminosity of the object, then the luminosity distance $d_L = \left(\frac{L}{4\pi\ell}\right)^{\frac{1}{2}}$ evolves as

$$d_L = d(1+z) = \frac{z+1}{\beta(\beta-2)}\Big(-\frac{\ln(1+z)}{\alpha}\Big)^{-1+\frac{2}{\beta}}. \tag{27}$$

Figure (7) shows the variation of the luminosity distance as a function of the exponent $0 < n < 1$.

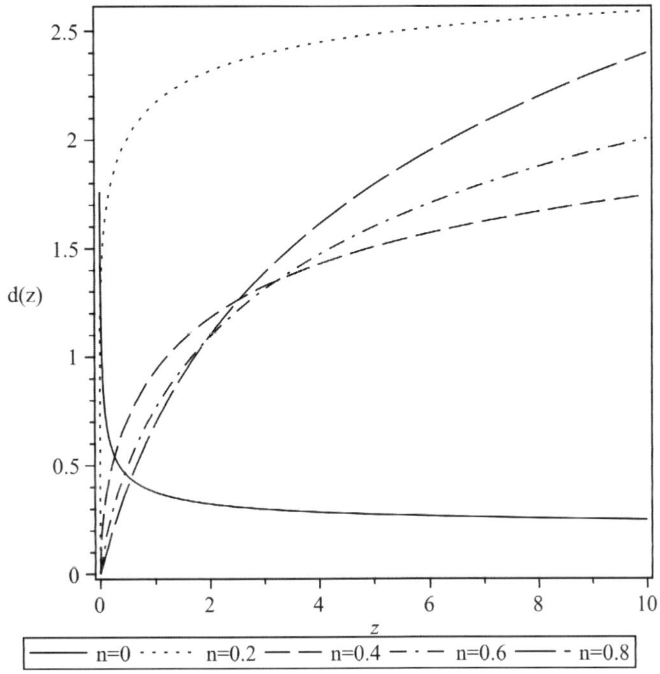

Figure 6. The behavior of the proper distance as a function of redshift for $\alpha = -1, 0 < n < 1$.

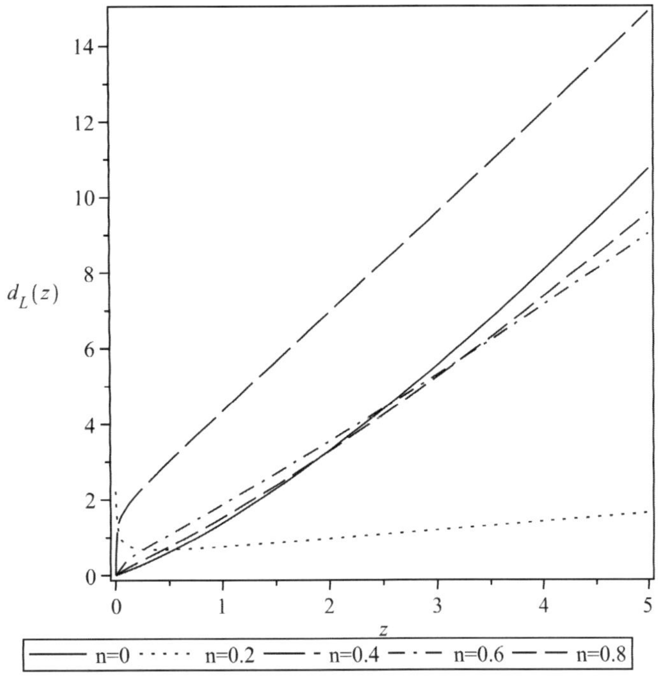

Figure 7. The behavior of the luminosity distance as a function of redshift for $\alpha = -1, 0 < n < 1$.

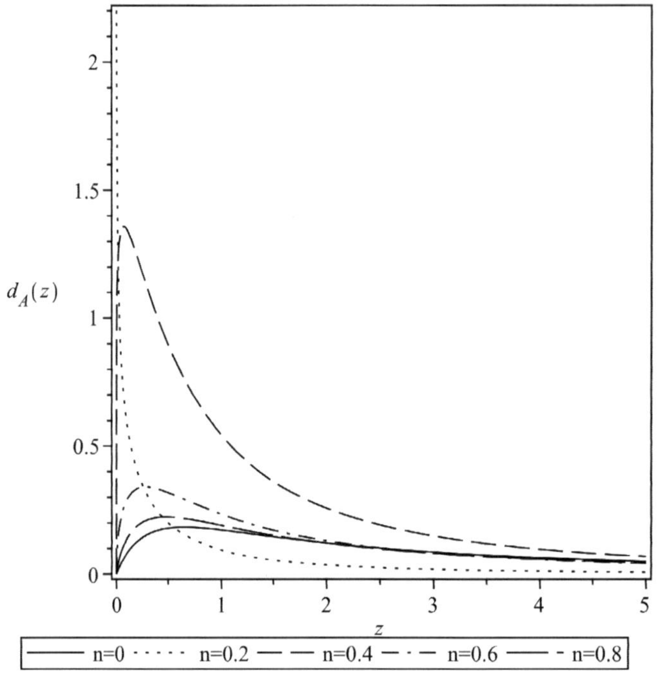

Figure 8. The behavior of the angular diameter as a function of redshift for $\alpha = -1, 0 < n < 1$.

2.6. Angular diameter

The angular diameter distance d_A is

$$d_A = d_L(1+z)^{-2} = \frac{1}{\beta(\beta-2)(1+z)}(-\frac{\ln(1+z)}{\alpha})^{-1+\frac{2}{\beta}}. \tag{28}$$

Figure-8 shows the variation of the angular diameter as a function of the exponent $0 < n < 1$.

2.7. State parameter of dark energy

In this subsection we discuss the dynamics of the dark energy. We know that in a flat FRW model with a perfect fluid of pressure p and energy density ρ, the barotropic index $w(t)$ is defined by

$$w(t) = -1 - \frac{\partial_t \ln(\rho(t))}{3H}, \tag{29}$$

which can be can be derived from the continuity equation. Using (9),(10) with (29) we obtain:

$$w(t) = -1 - \frac{1}{3\,(-1+n)\,c_n}\, np\chi^{\frac{n}{2(-1+n)}} t^{\frac{pn+2-2\,n}{2(-1+n)}}. \tag{30}$$

For simplicity we take $D = 2$, $\chi << 1$ and $G = \frac{1}{8\pi}$. Like the previous sections, we are interested in expressing w as a function of z. We substitute $\beta = \frac{2+(p-2)n}{2(1-n)}$ and from Eqs. (24) and (30) we have

$$w(t) \quad = \quad -1 - \frac{1}{3\,(-1+n)\,c_n}\, np\chi^{\frac{n}{2(-1+n)}} \left(\left[\frac{-\log(1+z)}{\alpha}\right]^{\frac{1}{\beta}}\right)^{\frac{pn+2-2\,n}{2(-1+n)}}.$$

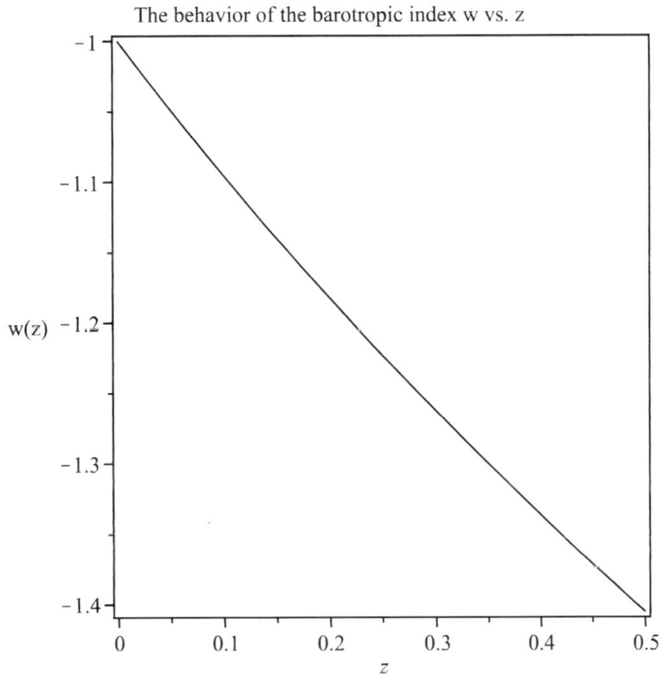

Figure 9. The behavior of the barotopic index w vs. z

Since $\frac{1}{\beta} \times \frac{pn+2-2n}{2(1-n)} = 1$, therefore the previous equation reads

$$w(t) = -1 + \frac{1}{3\alpha\,(-1+n)\,c_n} np\chi^{\frac{n}{2(1-n)}} \log(1+z),$$

Now using the values of ρ_n, b_n, c_n and Λ_n, we can write

$$w(z) = -1 - d_n \ln(1+z)\,, \tag{31}$$

where

$$d_n = \frac{2p}{\sqrt{3}}\Big(\frac{n^n\chi^{2-n}}{\Lambda_0(1-n)}\Big)^{\frac{1}{2(1-n)}}. \tag{32}$$

Also we observe from the previous figures, the model parameters do not depend on p significantly and since the fractal parameter $\chi << 1$ therefore we limited ourselves only to $p = 1$. From (30) it is obvious that the barotropic index crosses the dark energy limit $w(t) = -1$ for current era $z = 0$ several times for different values of χ and n. For instance, using typical values $\chi = 0.05, n = 0.5, \Lambda_0 = .01826$, (30) becomes

$$w(t) = -1 - \ln(1+z)\,. \tag{33}$$

Figure-9 shows the behavior of the barotopic index w against z. The pressure may be written as

$$p(z) = -p_n(\ln(1+z) + d_n \ln(1+z)^{\frac{2-n}{n}}), \tag{34}$$

where

$$p_n = \rho_n\Big(-\frac{\chi}{\alpha}\Big)^{\frac{1}{1-n}}. \tag{35}$$

2.8. *Consistency of the model $G(t) = \frac{G}{t}$ with observational constraints*

From observational data, as it was shown in [20], there is a secular decrease of the gravitational constant $G(t)$. This time rate of change is of order $\frac{\dot{G}}{G} = (-5.9 \pm 4.4) \times 10^{-14} yr^{-1}$ which was obtained from planetary data analysis. Thus, in our treatment on fractional cosmology focusing on the model $G(t) = \frac{G}{t}$, we can accept it as a reasonable model in good agreement with observational constraints [21].

Final remarks

This work was motivated by a more recent work on fractional-action cosmology with a periodic weight function in the action [1] which we extended using a power-law weight function. By doing so we obtained a varying gravitational coupling constant and also a time varying cosmological constant. For both of these 'constants', we used different ansatz to analyze the behavior of other parameters. We also modeled dark energy in this paradigm and obtained relevant cosmological parameters including distance parameters and statefinder parameters. The former helps in determining the nature of dark energy while the latter one classify different dark energy candidates.

Acknowledgments

The authors would like to thank R. A. El-Nabulsi for useful comments and valuable suggestions during this work. Also, we thank E. Pitjeva for pointing the note on secular decrease of the gravitational constant $G(t)$. M. Jamil would like to thank NUST for providing support to present this work at the III Italian-Pakistani Workshop on Relativistic Astrophysics in Italy.

References

[1] EL. Nabulsi R A 2010 *Commun. Theor. Phys* **54**, 16.
[2] Weyl H 1917 *Ann. Phys.* **54**, 117.
[3] Weyl H 1919 *Ann. Phys.* **59**, 101.
[4] Dirac P A M 1973 *Proc. R. Soc. Lond.* **333**, 403.
[5] Canuto V, Hsieh S H, Adams P J 1977 *Phy. Rev. Lett.* **39** 8.
[6] Bousso R 2000 *JHEP* **0011**, 038.
[7] Blake G M 1978 *Mon. Not. R. Astron. Soc.* **185**, 399.
[8] Peebles P J E, Ratra B 2003 *Rev. Mod. Phys.* **75**, 559.
[9] Eddington A 1931 *Proc. Cam. Phil. Soc.* **27** (1931)
[10] Stewart J *The Deadbeat Universe Lars Wÿahlin*, Colutron research, Boulder, Colorado, chap. 8, pp 104 (1931).
[11] Jordan P *Die Herkunft der Sterne* (Stuttgart : Wissenschaftliche Verllagsgesellschaft M.B.H., 1947)
[12] Shemi-zadeh V E 2002 gr-qc/0206084.
[13] Dirac P A M 1937 *Nature* **139**, 323.
[14] Barrow J D 2002 *The Constants of Nature*, (Pantheon Books, 2002).
[15] El-Nabulsi A R 2007 *Rommanian Rep. Phys.* **59**, 763.
[16] De Felice A, Tsujikawa S 2010 *Living Rev. Relativity* **13** ,3.
[17] Cai Y-F, Saridakis E N, Setare M R, Xia J-Q 2010 *Phys. Rep.* **493**, 1.
[18] Sahni V, Saini T D, Starobinsky A A, Alam U 2003 *JETP Lett.* **77**, 201.
[19] Jamil M, Debnath U 2011 *Int. J. Theor. Phys.* **50**, 1602.
[20] Pitjeva E V 2010 EPM ephemerides and relativity (in) Relativity in Fundamental Astronomy: Dynamics, Reference Frames and Data Analysis, Proceedings IAU Symposium No. 261, Klioner S A, Seidelmann P K, Soffel M H, (eds.) Cambridge University Press, Cambridge, pp. 170-178.
[21] http://iau-comm4.jpl.nasa.gov/plan-eph-data/index.html

Journal of Physics: Conference Series **354** (2012) 012009 doi:10.1088/1742-6596/354/1/012009

Limits on decaying dark energy density models from the CMB temperature-redshift relation

Philippe Jetzer and Crescenzo Tortora

Universität Zürich, Institut für Theoretische Physik, Winterthurerstrasse 190, CH-8057, Zürich, Switzerland

Abstract. We discuss the thermodynamic and dynamical properties of a variable dark energy model with density scaling as $\rho_x \propto (1+z)^m$, z being the redshift. These models lead to the creation/disruption of matter and radiation, which affect the cosmic evolution of both matter and radiation components in the Universe. In particular, we have studied the temperature-redshift relation of radiation, which has been constrained using a recent collection of cosmic microwave background (CMB) temperature measurements up to $z \sim 3$. We find that, within the uncertainties, the model is indistinguishable from a cosmological constant which does not exchange any particles with other components. Future observations, in particular measurements of CMB temperature at large redshift, will allow to give firmer bounds on the effective equation of state parameter $w_{\rm eff}$ for such types of dark energy models.

1. Introduction

The current standard cosmology model relies on the existence of two unknown dark components, the so called "dark matter" (DM) and "dark energy" (DE), which amount to $\sim 25\%$ and $\sim 70\%$ of the total energy budget in the Universe, respectively. According to several observations, the Universe is spatially flat and in an accelerated phase of its expansion [1, 2, 3, 4, 5]. DE, described as a cosmological constant Λ in its simplest form, is modelled by a fluid with a negative pressure, which is a fundamental ingredient to explain the actual accelerated expansion of the Universe.

Several models have been proposed to explain DE [6, 7, 8, 9, 10, 11, 12, 13]. An alternative consists to consider a phenomenological variable DE density with continuous creation/disruption of photons [14, 15, 16, 17, 18, 19] or matter [20, 21]. The DE might decay/grow slowly in the course of the cosmic evolution and thus provide the source/sink term for matter and radiation. Different such models have been discussed and strong constraints come from very accurate measurements of the cosmic microwave background (CMB) radiation and other typical cosmological probe.

CMB radiation is the best evidence for an expanding Universe starting from an initial high density state. Within the Friedmann-Robertson-Walker (FRW) models of the Universe the radiation, after decoupling, expands adiabatically and scales as $(1+z)$, z being the redshift [22]. If we assume that each component is not conserved, contrarily to the standard scenario, then depending on the decay mechanism of the DE, the created photons could lead to distortions in the Planck spectrum of the CMB, and change the evolution of its temperature. The chance to appreciate the deviation from the standard temperature evolution is given by the increasingly number of recent works collecting observations of CMB temperature both at low [23, 24] and higher redshifts [25].

Following the theoretical lines of [14, 15, 16, 17], in Jetzer et al. [18, 19] we have discussed a variable DE model $\Lambda(z) \propto (1+z)^m$ decaying into photons and DM particles. In particular we studied thermodynamical aspects in the case of a continuous photon creation, which implies a modified temperature redshift relation for the CMB. We have tested the predicted temperature evolution of radiation with some recent data on the CMB at higher redshift from both Sunyaev-Zel'dovich (SZ) effect and high-redshift QSO absorption lines as well as with an updated collection of data from different kind of observations, like distance moduli of Supernovae Ia, observations of the CMB anisotropy and the large-scale structure, together with observational Hubble parameter estimations [19]. A similar approach, however without considering photon creation and thus a modification in the CMB temperature evolution, has been discussed by Ma [21].

2. Theoretical framework

We assume a cosmological framework based on the usual Robertson-Walker (RW) metric element [22] and that the Universe contains three different components: a) a matter (both baryons and DM) fluid, with equation of state $p_m = 0$ (since $p_m << \rho_m$), b) a generalised fluid with pressure $p_\gamma = (\gamma - 1)\rho_\gamma$, where γ is a free parameter, which is set to 4/3 for a properly said radiation fluid, and c) a DE, x component, with pressure p_x and density ρ_x. The equation of state for the x component could assume a very general expression, but we limit ourselves to consider the simple linear relation $p_x = w_x \rho_x$. We will set any 'bare' cosmological constant Λ_0 equal to 0 [18]. With these components we get for the Einstein field equations [22]

$$8\pi G \rho_{tot} = 3\frac{\dot{R}^2}{R^2} + 3\frac{k}{R^2} , \tag{1}$$

$$8\pi G p_{tot} = -2\frac{\ddot{R}}{R} - \frac{\dot{R}^2}{R^2} - \frac{k}{R^2} , \tag{2}$$

where $p_{tot} = p_\gamma + p_x$ and $\rho_{tot} = \rho_m + \rho_\gamma + \rho_x$ are the total pressure and density, R is the scale factor, $k = 0, \pm 1$ is the curvature parameter and a dot means time derivative. Furthermore, we will assume that there is no curvature, thus $k = 0$ [3].

In the following we will adopt $w_x = -1$, however since we are assuming that the vacuum decays into radiation and massive particles, the effective equation of state w_{eff}, which is the measured quantity, can differ from -1.

Following [18, 19] (but see also [14, 16]) we set here the energy conservation equation for the different fluids. The fluid as whole, verifies the Bianchi identity, which means that energy and momentum are locally conserved: $\nabla_\mu T^{\mu\nu} = 0$, with $T^{\mu\nu}$ the stress-energy tensor

$$T^{\mu\nu} = (\rho_{tot} + p_{tot}) u^\mu u^\nu - p_{tot} g^{\mu\nu} \tag{3}$$

and u^μ being the four velocity. After easy calculations, this conservation law reads

$$\dot{\rho}_{tot} + 3(\rho_{tot} + p_{tot}) H = 0 , \tag{4}$$

where $H = \dot{R}/R$ is the Hubble parameter. In the standard approach, each component is conserved, thus Eq. (4) holds for all the fluid components, but here we will suppose that each component will exchange energy with each other. In particular, for matter, radiation and DE we impose the following relations (see also [26])

$$\dot{\rho}_m + 3\rho_m H = (1-\epsilon)\, C_x , \tag{5}$$

$$\dot{\rho}_\gamma + 3\gamma\rho_\gamma H = \epsilon\, C_x , \tag{6}$$

$$\dot{\rho}_x + 3(p_x + \rho_x)H = -C_x \ , \tag{7}$$

where we assume that both the matter and the radiation fluids exchange energy with the DE as parameterized by C_x and ϵ. In particular, C_x depends on the DE and, indeed, acts as a source/sink term for the fluids energy. Evidently, if no interaction between the different components exists, then C_x is null and the standard picture is recovered. Moreover, if $\epsilon = 0$ ($\epsilon = 1$), then the DE exchanges all the energy with matter (radiation). C_x can describe different physical situations such as, for instance, a thermogravitational quantum creation theory [15] or a quintessence scalar field cosmology [7]. As we will discuss later on, ϵ has to be very small (i.e., $\epsilon \ll 1$), otherwise the radiation density would become much too big, contrary to present values. As an order of magnitude estimate we expect $\epsilon \simeq \frac{p_\gamma+\rho_\gamma}{\rho_m}$, for which indeed $\epsilon \ll 1$, since $p_\gamma + \rho_\gamma \ll \rho_m$.

Adopting as mentioned above the relation $p_x = -\rho_x$ and defining $\rho_x = \Lambda(t)/(8\pi G)$, from Eq. (7) we obtain

$$C_x = -\frac{\dot{\Lambda}(t)}{8\pi G} \ . \tag{8}$$

We assume a power law model for the $\Lambda(t)$ function, $\Lambda(t) = B(R/R_0)^{-m}$, or equivalently in terms of redshift, $\Lambda(z) = B(1+z)^m$, where B is a constant. The value of this constant is $B = 3H_0^2(1-\Omega_{\rm m0})$, which can be found using Eq. (1) at the present epoch, assuming that today $\rho_\gamma \ll \rho_m, \rho_x$. Thus, the density evolution for the x component is given by

$$\rho_x/\rho_{crit} = (1-\Omega_{\rm m0})(1+z)^m, \tag{9}$$

where we have defined as $\rho_{crit} = \frac{3H_0}{8\pi G}$ the present critical density of the Universe. If m is positive, then the DE slowly decreases as a function of the cosmic time, whereas if m is negative the inverse process happens.

From Eqs. (5) and (6) we derive the evolution laws for matter and radiation,

$$\begin{aligned}\rho_m/\rho_{crit} &= \Omega_{\rm m0}(1+z)^3 \\ &\quad - (1-\epsilon)\frac{m(1-\Omega_{\rm m0})}{m-3}[(1+z)^m - (1+z)^3] \ ,\end{aligned} \tag{10}$$

$$\begin{aligned}\rho_\gamma/\rho_{crit} &= \Omega_{\gamma0}(1+z)^{3\gamma} \\ &\quad - \epsilon\frac{m(1-\Omega_{\rm m0})}{m-3\gamma}[(1+z)^m - (1+z)^{3\gamma}] \ ,\end{aligned} \tag{11}$$

where $\Omega_{\rm m0}$ and $\Omega_{\gamma0}$ are the matter and radiation energy densities at $z = 0$, respectively.

Since we are interested in the evolution of the radiation temperature, it is useful to discuss the extreme case when only photons enter in the process (i.e. $\epsilon = 1$). Then for the matter density in Eq. (10) the usual evolution $\propto (1+z)^3$ holds, while the radiation besides the usual term $\Omega_{\gamma0}(1+z)^{3\gamma}$ has also a perturbative term depending on m. Since today [1] $\Omega_{\gamma0} \sim 5\times10^{-5}$ it turns out that m has to be extremely small $\lesssim 10^{-4}$. Therefore, unless m is extremely small or vanishing, DE has to decay mainly in matter with possibly some photons as well. Thus the condition $\epsilon \ll 1$ has to hold.

[1] The present radiation density is the only cosmological parameter accurately measured. The radiation density is dominated by the energy in the cosmic microwave background (CMB), and the COBE satellite FIRAS experiment determined its temperature to be $T = 2.725 \pm 0.001\,K$ [27], corresponding to $\Omega_{\gamma0} \sim 5\times10^{-5}$.

2.1. Hubble and deceleration parameter

Due to the very small value of $\Omega_{\gamma 0}$ it follows that the evolution of the Universe is essentially driven by the DE and DM, therefore, from Eq. (1) the following law for the Hubble parameter holds

$$\begin{aligned} H(z) &\simeq \frac{8\pi G}{3}(\rho_m + \rho_x) \\ &= H_0 \left[\frac{3(1-\Omega_{\mathrm{m}0})}{3-m}(1+z)^m + \frac{(3\Omega_{\mathrm{m}0} - m)}{3-m}(1+z)^3 \right]^{1/2} , \end{aligned} \tag{12}$$

which is obviously the same expression found in Ma [21].

Recasting Eq. (7), it is possible to write

$$\dot{\rho}_x + 3H(p_x + \rho_x + \frac{C_x}{3H}) = 0 , \tag{13}$$

which shows that the term C_x contributes to an effective pressure

$$p_{\mathrm{eff}} = p_x + \frac{C_x}{3H} = -\rho_x + \frac{C_x}{3H} . \tag{14}$$

Therefore, we get an equivalent effective DE equation of state w_{eff} [21]

$$w_{\mathrm{eff}} = \frac{p_{\mathrm{eff}}}{\rho_x} = \frac{m}{3} - 1. \tag{15}$$

If $m > 0$ then we have $w_{\mathrm{eff}} > -1$, i.e. our model is quintessence-like [6, 7, 12], while we have a phantom-like [9] model when m is negative and $w_{\mathrm{eff}} < -1$. Another interesting quantity is the deceleration parameter, which can be written as

$$\begin{aligned} q(z) = -\frac{\ddot{R}R}{\dot{R}^2} = \\ \frac{(1+z)^3(m-3\Omega_{\mathrm{m}0}) + 3(m-2)(1+z)^m(\Omega_{\mathrm{m}0}-1)}{2(1+z)^3(m-3\Omega_{\mathrm{m}0}) + 6(1+z)^m(\Omega_{\mathrm{m}0}-1)} . \end{aligned} \tag{16}$$

Imposing that $q(z) = 0$, we can determine the *transition redshift*, i.e. the redshift at which the Universe changed from a deceleration to an acceleration phase, which is given by

$$z_T = \left(\frac{3(2-m)(1-\Omega_{\mathrm{m}0})}{3\Omega_{\mathrm{m}0} - m} \right)^{\frac{1}{3-m}} - 1 . \tag{17}$$

From this result we see that the larger m is, the earlier the Universe changes from deceleration to acceleration.

2.2. Thermodynamical aspects and CMB temperature evolution

We follow here the approach outlined in Lima [14], where he defines a current as $N^\alpha = nu^\alpha$ with n being the particle number density of the photons or of the DM particles. Indeed, there is a current for each of these components. Due to the decaying vacuum the current satisfies the following balance equation (one for each component)

$$\dot{n}_i + 3n_i H = \psi_i , \tag{18}$$

with $i = \gamma$ or DM (γ for the photons and DM for the dark matter) and ψ_i is the corresponding particle source. For decaying vacuum models $\psi_\gamma + \psi_{DM}$ is positive and related to the rate of change of ρ_x. We can also define an entropy current of the form

$$S^\alpha = \sum_i n_i \sigma_i u_i^\alpha \ , \tag{19}$$

where σ_i is the specific entropy per particle (photons, DM and in principle also DE). If the DE ρ_x is constant the above entropy current is conserved. The existence of a non equilibrium decay process of the vacuum implies $S^\alpha_{;\alpha} \geq 0$, thus an increase of the entropy as a consequence of the second law of thermodynamics. In principle the second law should be applied to the system as a whole, thus including the vacuum component [26]. Assuming that the vacuum is like a condensate with zero chemical potential μ_{vac} it follows from Euler's relation

$$\mu_{vac} = \frac{\rho_x + p_x}{n} - T\sigma_{vac} \ , \tag{20}$$

provided $w_x = -1$, that $\sigma_{vac} = 0$ and thus its contribution to the entropy current vanishes. Given our assumptions the vacuum plays the role of a condensate carrying no entropy.

For instance, for quintessence models in the limit where the scalar field does not depend on time, and thus its time derivative vanishes, one gets $w_x = -1$. In the later stages of the Universe the time dependence is possibly very weak so that $w_x = -1$ holds up to small corrections.

The equation for the particle number density of radiation component is given by Eq. (18) with $i = \gamma$. Using Gibbs law and well-known thermodynamic identities, following the derivation given in the paper by Lima et al. [16], one gets (see also [14, 18])

$$\frac{\dot{T}}{T} = \left(\frac{\partial p_\gamma}{\partial \rho_\gamma}\right)_n \frac{\dot{n}_\gamma}{n_\gamma} - \frac{\psi_\gamma}{n_\gamma T \left(\frac{\partial \rho_\gamma}{\partial T}\right)_n} \left[p_\gamma + \rho_\gamma - \frac{n_\gamma \epsilon C_x}{\psi_\gamma}\right] \ . \tag{21}$$

To get a black-body spectrum the second term in brackets in Eq. (21) has to vanish, thus

$$\epsilon C_x = \frac{\psi_\gamma}{n_\gamma}\left[p_\gamma + \rho_\gamma\right] \ . \tag{22}$$

Thus, Eq. (21) becomes

$$\frac{\dot{T}}{T} = \left(\frac{\partial p_\gamma}{\partial \rho \gamma}\right)_n \frac{\dot{n}_\gamma}{n_\gamma} \ . \tag{23}$$

With $\left(\frac{\partial p_\gamma}{\partial \rho_\gamma}\right)_n = (\gamma - 1)$ one obtains

$$\frac{\dot{T}}{T} = (\gamma - 1)\frac{\dot{n}_\gamma}{n_\gamma} \ . \tag{24}$$

Using the equation for the particle number conservation Eq. (18) into Eq. (24) leads to

$$\frac{\dot{T}}{T} = (\gamma - 1)\left[\frac{\psi_\gamma}{n_\gamma} - 3H\right] \ . \tag{25}$$

With Eqs. (22) and (25) we get

$$\frac{\dot{T}}{T} = (\gamma - 1)\left[-\frac{\epsilon\,\dot{\Lambda}}{8\pi G(p_\gamma + \rho_\gamma)} - 3H\right] \ . \tag{26}$$

Now, following the previous discussion on ϵ and aiming to be very general, we set $\epsilon = \frac{\rho_\gamma + p_\gamma}{\rho_m}\tilde{\epsilon}$, where $\tilde{\epsilon}$ is a new parameter and insert it into Eq. (26). Taking the sum of Eqs. (1) and (2) we find

$$8\pi G(\rho_{tot} + p_{tot}) \simeq 8\pi G\rho_m = 2\frac{\dot{R}^2}{R^2} - 2\frac{\ddot{R}}{R} = -2\dot{H} \ . \tag{27}$$

Finally, we obtain the expression

$$\frac{\dot{T}}{T} = (\gamma - 1)\left[\frac{\dot{\Lambda}\tilde{\epsilon}}{2\dot{H}} - 3H\right] , \tag{28}$$

which we can integrate

$$\int_{t_1}^{t_0} \frac{\dot{T}}{T} dt = (\gamma - 1)\int_{t_1}^{t_0}\left[\frac{\dot{\Lambda}\tilde{\epsilon}}{2\dot{H}} - 3H\right] dt , \tag{29}$$

where t_0 denotes the present time and t_1 some far instant in the past. Indeed, if $\dot{\Lambda}$ vanishes and $\gamma = 4/3$ one gets the usual dependence $T(t) = \frac{R(t_1)T(t_1)}{R(t)}$ for a radiation fluid. To carry out the integration of the first term on the right hand side it is useful to perform a change of variable from t to z and accordingly $\frac{dt}{dz} = \frac{-1}{H(1+z)}$. This way we get (with z_1 corresponding to the time t_1 and $z_0 = 0$ corresponding to t_0 present time)

$$\begin{aligned} ln\frac{T(z=0)}{T(z_1)} + 3(\gamma - 1)ln\frac{R(z=0)}{R(z_1)} &= \\ \frac{(\gamma-1)}{2}\int_0^{z_1} \frac{\Lambda'\tilde{\epsilon}}{H'H(1+z)}dz , \end{aligned} \tag{30}$$

where $'$ denotes derivative with respect to z.

As next we insert $H(z)$ and its derivative as taken from Eq. (12) into Eq. (30) and integrate it, to get (setting $z_1 = z$)

$$T(z) = T_0\left(\frac{R_0}{R(z)}\right)^{3(\gamma-1)} exp\left(\frac{B(1-\gamma)\tilde{\epsilon}}{3H_0^2(\Omega_{\rm m0} - 1)}A\right) , \tag{31}$$

where

$$\begin{aligned} A = ln((m - 3\Omega_{\rm m0}) + m(1+z)^{m-3}(\Omega_{\rm m0} - 1)) \\ -ln((m-3)\Omega_{\rm m0})] \ . \end{aligned} \tag{32}$$

We can also write Eq. (31) as

$$\begin{aligned} T(z) = T_0(1+z)^{3(\gamma-1)} \\ \times\left(\frac{(m - 3\Omega_{\rm m0}) + m(1+z)^{m-3}(\Omega_{\rm m0} - 1)}{(m-3)\Omega_{\rm m0}}\right)^{\tilde{\epsilon}(\gamma-1)} . \end{aligned} \tag{33}$$

We inserted in the exponent of Eq. (31) the explicit form of B, thus getting as exponent in the above Eq. $\tilde{\epsilon}(\gamma - 1)$. Hereafter, we will set $\tilde{\epsilon} = 1$. Clearly $\tilde{\epsilon}$ and m are not independent, we checked using the temperature redshift data that if $\tilde{\epsilon}$ is bigger ($\sim$ 10 or more), then m has to be extremely small consistently with what mentioned above (as it would lead to a too high production of photons in the DE decay). On the other hand, if $\tilde{\epsilon}$ gets smaller (e.g., ~ 0.1) m gets bigger (~ 0.2) and accordingly w_{eff}, moreover m would be poorly constrained, since the uncertainties would then be very high. But from the other data (without the the temperature ones) there are already stringent limits on m and thus this way one could get lower limits on $\tilde{\epsilon}$, under the assumption that DE decays also in photons.

Notice that for $z = 0$ we have $T(0) = T_0$, whereas for $m = 0$ the expression in the parenthesis is equal to 1 and thus $T(z) = T_0(1+z)^{3(\gamma-1)}$, which for the canonical value of $\gamma = 4/3$ reduces to the standard expression.

3. Results

To test the temperature evolution for the radiation component, we rely on the CMB temperatures derived from the absorption lines of high redshift systems and the ones from SZ effect in clusters of galaxies (we will collectively quote as T_{CMB}, hereafter). At high redshift the CMB temperature is recovered from the excitation of interstellar atomic or molecular species that have transition energies in the sub-millimetre range and can be excited by CMB photons. When the relative population of the different energy levels are in radiative equilibrium with the CMB radiation, the excitation temperature of the species equals that of the black-body radiation at that redshift, providing one of the best tools for determining the black-body temperature of the CMB in the distant Universe [28, 29, 30, 31, 32, 33, 34, 35]. We used a sample of 9 QSO absorption measurements [19, 25]. In summary we have 4 data points from the analysis of the fine structure of atomic carbon (AC) and 5 measurements based on the rotational excitation of CO molecules (CO) [25].

At lower redshift we use the measurements from the SZ effect. Thus, spectral measurements of galaxy clusters at different frequency bands yield independent intensity ratios for each cluster. The combinations of these measured ratios permit to extract the cosmic microwave background radiation (see Fabbri et al. [36]). We relied on the data compilation in Luzzi et al. [24], which have analyzed the results of multifrequency SZ measurements toward several clusters from 5 telescopes (BIMA, OVRO, SUZI II, SCUBA and MITO).

We derive the observed T_{CMB} from the theoretical expression T_{th}, which we have derived in Eq. (33). We set $T_0 = 2.725$ K, which is quite well determined in the literature [37], and the matter density $\Omega_{\rm m0} = 0.273$ to the value inferred in Komatsu et al. [38]. If we take $\gamma = 4/3$, then we find $m = 0.03^{+0.08}_{-0.09}$ [19], which is lower than the estimated value of $m = 0.09 \pm 0.10$ in [18], but fully consistent within uncertainties, and also pretty consistent with $m = 0$. In Fig. 1 the temperature measurements (together with some upper limits) are shown, and our best fitted result is plotted and compared with the $m = -0.09$ result in Ma [21]. The value we have found corresponds to an effective equation of state $w_{\rm eff} = -0.99 \pm 0.03$, consistent with $w_{\rm eff} = -1$, and the transition redshift is $z_T = 0.78 \pm 0.08$. In order to check the impact of redshift distribution we separate the data in two redshift bins with $z < 0.6$ (SZ data only) and $z \geq 0.6$ (QSO absorption lines only), finding the best fitted values $m = 0.12^{+0.12}_{-0.13}$ and $-0.05^{+0.12}_{-0.14}$, respectively. Although the uncertainties are very high, it seems to emerge a mild trend with lower redshift data preferring a DE decaying into matter and radiation, while data at $z > 0.6$ point to an opposite behavior. These results could be interpreted in a different way, in fact, the differences found could be due not to the different redshift coverage, but to some particular biases in the two kind of observations, SZ vs QSO absorption lines. An indication of this suggestion comes if we divide the sample in three subsamples: 1) SZ data, 2) the data from the analysis of the fine structure of atomic carbon (AC), and 3) the measurements based on the rotational excitation of CO molecules in [25]. If we fit the model to the combined SZ+AC and SZ+CO samples we find $m = 0.11^{+0.11}_{-0.11}$ and $m = 0.03^{+0.09}_{-0.10}$, respectively. Because of the larger measurement errors, the AC data affect very little the results when only SZ measurements are adopted, and the result for the second sample shows that SZ and CO data mainly constrain m, and AC simply gives a tiny reduction on the errors.

Adopting a constant value for the ratio $\psi_\gamma / 3n_\gamma H = \beta$ Lima et al. [16] have found the simple

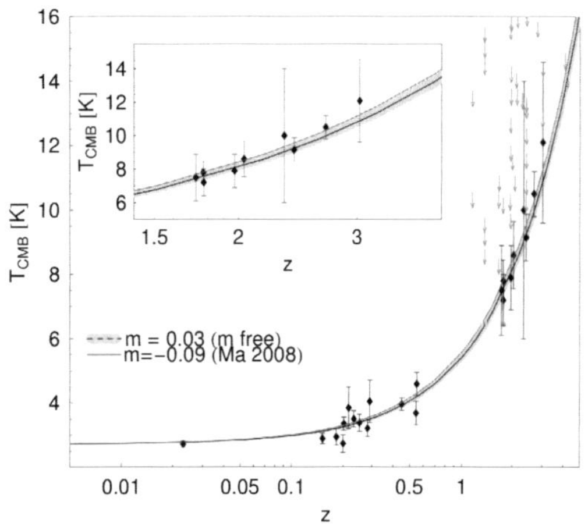

Figure 1. Cosmic microwave background temperature as a function of the redshift. The black points with bars are the full collection of measurements from Luzzi et al. [24] and Noterdaeme et al. [25]. The gray arrows represent the upper limits derived from the analysis of atomic carbon (see [25] for details). The black line is the best fit result ($m = 0.03$), while the gray region is the 1σ uncertainty. The red line is the best fit recovered from Ma [21] (see the legend). The inset panel show a magnified vision of the higher redshift region of the plot.

relation,

$$T(z) = T_0(1+z)^{1-\beta} \, . \tag{34}$$

If we fit all the sample we find $\beta = -0.002 \pm 0.03$, while for the low and high redshift subsamples we have $\beta = 0.06 \pm 0.08$ and -0.01 ± 0.03, respectively. These results are qualitatively consistent with what found in Noterdaeme et al. [25], and points to a similar trend as the one discussed above.

If γ is left free to change, for the best fitted value it turns out that $\gamma > 4/3$ and m is systematically more positive. We find $\gamma = 1.35^{+0.03}_{-0.03}$ and $m = 0.25^{+0.23}_{-0.17}$, which corresponds to an effective equation of state $w_{\rm eff} = -0.92 \pm 0.07$ and the transition redshift is $z_T = 1.1 \pm 0.6$. When the two subsamples are adopted, wide confidence contours are found, particularly for the $z \geq 0.6$ sample, for which the contours at very low m are not closed. We obtain $\gamma = 1.3^{+0.2}_{-0.1}$ and $1.26^{+0.01}_{-0.01}$, while $m = 0.8^{+0.1}_{-0.3}$ and $0.6^{+0.1}_{-1.0}$, respectively for the low and high-z samples.

We notice that if DE does not decay into radiation (corresponding to $\epsilon = \tilde{\epsilon} = 0$ and thus m is no longer constrained) then the CMB temperature will scale in the standard way. Clearly, this would imply that if DE decays, this has to be into DM only. On the other hand a deviation of the CMB temperature from the standard scaling could be interpreted as DE decaying also into radiation. In which case with Eq. (33) one can determine either m or $\tilde{\epsilon}$ and thus get some insights on the decay mode of DE into radiation. Future data on the CMB temperature will allow to shed light on this important issue.

In [19] we have combined the CMB temperature data with a set of other independent measurements such as Supernovae Ia distance moduli, CMB anysotropy, baryon acoustic oscillation and observational data for the Hubble parameter. When combining all the data sets we found values for m, and thus $w_{\rm eff}$ consistent with the ones obtained when using only CMB temperature data.

C. Tortora was supported by the Swiss National Science Foundation.

[1] S. Perlmutter et al., Astrophys.J. **517**, 565 (1999)
[2] A.G. Reiss et al., Astron.J. **116**, 1009 (1998)
[3] P. de Bernardis et al., Nature **404**, 955 (2000)
[4] D.N. Spergel et al., Astrophys. J. Suppl. **170**, 377 (2007)
[5] R. Caldwell & M. Kamionkowski, Ann. Rev. Nucl. Part. Sci. **59**, 397 (2009)
[6] P.J.E. Peebles & B. Ratra, Astrophys. J. Lett. **325**, L17 (1988)
[7] B. Ratra & P.J.E. Peebles, Phys. Rev. D **37**, 3406 (1988)
[8] V. Sahni & A.A. Starobinsky, Int. J. Mod. Phys. **D9** 373 (2000)
[9] R.R. Caldwell, Phys. Lett. B **545**, 23 (2002)
[10] T. Padmanabhan, Phys. Rep. **380**, 235 (2003)
[11] P. J. E. Peebles and B. Rathra, Rev. Mod. Phys. **75**, 559 (2003)
[12] M. Demianski, E. Piedipalumbo, C. Rubano, C. Tortora, A&A **431** 27D (2005)
[13] V. F. Cardone, C. Tortora, A. Troisi, and S. Capozziello, Phys. Rev. **D73**, 043508 (2006)
[14] J.A.S. Lima, Phys. Rev. D. 54, 2571 (1996)
[15] J.A.S. Lima and J.S. Alcaniz, Astron. and Astrophys. **348**, 1 (1999)
[16] J.A.S. Lima, A.I. Silva and S.M. Viegas, Mon. Not. R. Astron. Soc. 312, 747 (2000)
[17] D. Puy, Astron. and Astrophys. **422**, 1 (2004)
[18] P. Jetzer, D. Puy, M. Signore, C. Tortora, Gen. Relativ. Gravit., **43**, 1083 (2011)
[19] P. Jetzer & C. Tortora, Phys. Rev **D84**, 043517 (2011)
[20] W. Chen, Y.S. Wu, Phys. Rev. D 41 (1990) 695; W. Chen, Y.S. Wu, Phys. Rev. D 45 (1992) 4728, Erratum.
[21] Y. Ma, Nucl. Phys. B **804**, 262 (2008)
[22] S. Weinberg, *Gravitation and Cosmology: Principles and Applications of the General Theory of Relativity* (Wiley, New York, 1972!.
[23] E.S. Battistelli et al., Astrophys. J. **580**, L101 (2002)
[24] G. Luzzi, M. Shimon, L. Lamagna, Y. Rephaeli, M. De Petris, A. Conte, S. De Gregori and E. Battistelli, Astrophys. J. **705**, 1122 (2009)
[25] P. Noterdaeme, P. Petitjean, R. Srianand, C. Ledoux, and S. López, A&A **526**, L7 (2011)
[26] M. Jamil, E.N. Saridakis and M.R. Setare, Phys. Rev. **D 81**, 023007 (2010)
[27] J.C. Mather et al., Astrophys. J. 512, 511 (1999).
[28] J. Ge, J. Bechtold and J. Black, Astrophys. J **474**, 67 (1997)
[29] R. Srianand, P. Petijean and C. Ledoux, Nature **408**, 931 (2000)
[30] P. Molaro, S. Levshakov, M. Dessauges-Zavadsky and S. D'Odorico, Astron. and Astrophys. **381**, L64 (2002)
[31] D. Puy, G. Alecian, J. Leorat, J. Lebourlot and G. Pineau des Forets, Astron. and Astrophys. **267**, 337 (1993)
[32] D. Galli and F. Palla, Astron. and Astrophys. **335**, 403 (1998)
[33] P. Stancil, S. Lepp and A. Dalgarno, Astrophys. J. **509**, 1 (1998)
[34] J. Cui, J. Bechtold, J. Ge and D. Meyer, Astrophys. J. **633**, 649 (2005)
[35] R. Srianand, P. Noterdaeme, C. Ledoux and P. Petijean, Astron. and Astrophys. **482**, L39 (2008)
[36] R. Fabbri, F. Melchiorri and V. Natale, Astrophys. Sp. Sci. **59**, 223 (1978)
[37] J.C. Mather et al., Astrophys. J. **512**, 511 (1999)
[38] E. Komatsu et al., Astrophys. J. Suppl. **180**, 330 (2009)

Journal of Physics: Conference Series **354** (2012) 012010 doi:10.1088/1742-6596/354/1/012010

Searching for continuous gravitational wave signals using LIGO and Virgo detectors

Paola Leaci (for the LIGO Scientific Collaboration and the Virgo Collaboration)

Max-Planck-Institut für Gravitationsphysik, Albert-Einstein-Institut, 14476 Golm, Germany

E-mail: paola.leaci@aei.mpg.de

Abstract. Direct and unequivocal detection of gravitational waves represents a great challenge of contemporary physics and astrophysics. A worldwide effort is currently operating towards this direction, building ever sensitive detectors, improving the modelling of gravitational wave sources and employing ever more sophisticated and powerful data analysis techniques. In this paper we review the current status of LIGO and Virgo ground based interferometric detectors and some data analysis tools used in the continuous wave searches to extract the faint gravitational signals from the interferometric noise data. Moreover we discuss also relevant results from recent continuous wave searches.

(Some figures in this article are in colour only in the electronic version)

1. Introduction

Gravitational waves, predicted to exist by Einstein's General Theory of Relativity [1], are ripples in space-time propagating at light speed and produced by non-axially symmetric mass accelerations, by analogy with electric charges in any accelerated motion that emit electromagnetic waves (travelling at light speed). However, gravitational waves are quite different from electromagnetic waves. They are both transverse waves, but gravitational waves are characterized by two polarization states, denoted as "+" and "×", that differ by a rotation of 45 degrees around the propagation axis, demonstrating the quadrupolar (spin-2) nature of the gravitational radiation. On the contrary, the two polarization states of electromagnetic waves differ by a rotation of 90 degrees, reflecting thus the dipolar (spin-1) characteristics of the electromagnetic radiation. The emission mechanisms are also quite different: gravitational waves result from the coherent emission from bulk motions of energy, while electromagnetic waves result from an incoherent superposition of waves from molecules, atoms and particles. The amount of energy radiated as gravitational waves by any mechanical system constructed by man is so small that it will probably never be observed. For this reason we hope to observe gravitational radiation emitted by sources at astrophysical distances. Indeed, even though gravitational waves have not yet been directly detected by any detectors, a very strong indirect proof of their existence was given by the observation of the binary pulsar PSR B1913+16, discovered in 1974 by the astronomers R. Hulse and J. Taylor [2]. Using the giant radio telescope at the Arecibo Observatory of Puerto Rico [3] to search systematically for pulsars, they observed that a particular pulsar (PSR B1913+16) was changing its motion rapidly and that the variation in pulse rate was caused by the changing Doppler effect. The decrease of the orbital period of such a pulsar around its companion could only be explained if angular momentum and energy were carried away from this system by gravitational waves.

Because of the extreme weakness of the interaction of gravitational radiation with matter, gravitational waves travel almost undisturbed from astrophysical sources to Earth, without being scattered or absorbed by interstellar dust and debris, carrying thus astronomical information which electromagnetic waves do not carry. An analysis of gravitational radiation would provide information of great value about the inaccessible and remote locations of the cosmos. It would tell us something about the behaviour of space-time and matter under the most extreme conditions, and it would also provide a check on Einstein's General Theory of Relativity. The detection of gravitational waves will help us to understand the dynamics of large-scale events in the Universe, like the death of whole stars, the explosion of quasars, the birth and the collisions of Black Holes (BHs for short; for more details see for instance Ref. [4] and references therein).

Nowadays laser interferometry is the basis of the most sensitive gravitational wave detectors, such as the LIGO (Laser Interferometer Gravitational wave Observatory) and Virgo detectors [5, 6, 7], whose current performances are concisely discussed in the next section.

The remainder of the paper is organized as follows. In Sec. 2, 3 and 4, respectively, we briefly overview the current gravitational wave detectors and their sources, with particular attention devoted to a short discussion of the continuous wave signal. Section 5 is devoted to describe some of the main methodologies used in the searches for continuous gravitational waves, highlighting the most recent results obtained analyzing LIGO and Virgo data. A summary is finally reported in Sec. 6.

2. Gravitational Wave Detectors

The pioneer of gravitational wave detection was Joseph Weber in the early 1960s, who developed the first resonant mass detector and later also investigated laser interferometry [8]. From that date until today, the experiments that aim at the detection of gravitational radiation, planned in laboratories throughout the world, are in continuous progress [9, 10, 11, 12, 13, 14, 15, 16, 7, 17, 18, 5]. However, in this paper we consider only the efforts related to ground-based interferometric detectors.

The current world-wide network of gravitational wave detectors consists of the following Michelson-type kilometer-scale laser interferometers:

(i) the French-Italian Virgo detector, at Cascina (Pisa, Italy), with 3 km arm length [7];

(ii) the German-British GEO experiment, near Hannover (Germany), with an arm length of 600 m [17];

(iii) the Japanese TAMA, located in Tokyo (Japan), with 300 m arm length [18];

(iv) the American LIGO project [5], that consists of three detectors working in unison; one at Livingston (Louisiana, USA), with an arm length of 4 km (LLO) and two in the same vacuum container at Hanford (Washington, USA), with an arm length of 4 km (LHO1) and 2 km (LHO2), respectively.

In 2007 LIGO and Virgo achieved their design sensitivities over a wide frequency range and today the performance of such interferometers is improved even more, as can be noticed in Fig. 1, where the noise spectra of such detectors is plotted versus the frequency. LIGO reached its design sensitivity with the fifth science run [S5 in short, started on 2005 November 4 (14) at LHO1 (LLO) and ended on 2007 October 1], but the sensitivity has continued to improve with time. In fact, with respect to S5, the sensitivity curve of a recently completed run, S6 (started on 2009 July 7 and ended on 2010 October 20), has a factor of 2 improvement above 300 Hz, as depicted in Fig. 1. Moreover, Virgo 2009 sensitivity measurements show a much better sensitivity than LIGO below ~ 40 Hz (see Fig. 1).

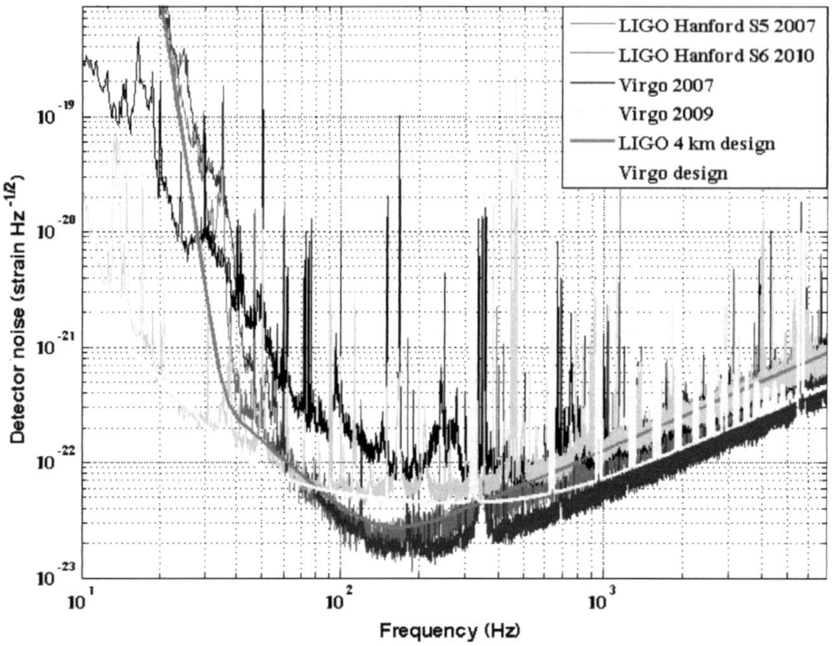

Figure 1. (color online). Strain sensitivity curves of the present gravitational wave interferometric detectors. In 2007 LIGO and Virgo reached their goal sensitivities in a wide frequency interval.

The upgrade of LIGO and Virgo interferometers for their Advanced stage is currently underway, with the goal of improving the current strain sensitivity by a factor ten, with a thousandfold increase in the observable volume of space. In 2015 such interferometers will be operational and will gradually improve their sensitivity. At their target sensitivity, several gravitational wave events per year should be detected, opening thus the gravitational wave astronomy era. A future project for an interferometer of comparable sensitivity, LCGT (Large Scale Cryogenic Gravitational Wave Telescope), is going to be built in Japan [19].

Detectors of third generation, such as Einstein telescope, are currently in design phase [20].

Moreover, space-borne detectors, such as LISA (Laser Interferometer Space Antenna) [21] and DECIGO/BBO (DECi-hertz Interferometer Gravitational wave Observatory and Big Bang Observer) [22], are designed to probe the 0.03 mHz to 0.1 Hz regime, bringing thus relevant astrophysical information at low frequencies and complementing the ground-based detectors.

In order to increase the baseline, it would be quite convenient to have a detector far away and out of the plane of other detectors in the USA and Europe. This would have a tremendous scientific impact and several advantages, such as to improve the ability to identify exactly where gravitational wave signals come from. The LIGO team is currently investigating some possibilities towards this direction.

3. Gravitational Wave Sources

Different types of gravitational wave sources are expected to be observed by ground-based detectors. It is well-known that coalescence of compact objects constitutes an interesting source of high-frequency gravitational waves [23]. In particular, the coalescence of NS-NS (NS, neutron star), NS-BH, BH-BH binary systems are expected to emit gravitational radiation in the kHz range [24]. Such kind of sources are referred to as *burst sources*, such as supernovae explosions,

Journal of Physics: Conference Series **354** (2012) 012010 doi:10.1088/1742-6596/354/1/012010

whose signals last for a very short amount of time, between a few milli-seconds and a few minutes.

A *stochastic background* of gravitational waves, either of cosmological or astrophysical origin, is also envisaged to exist. This consists of a random accumulation of signals from thousands or millions of individual sources. Last, but not least, another class of gravitational wave sources is represented by rapidly rotating non-axisymmetric NSs, that are predicted to emit continuously a weak sinusoidal signal. In the next sections we limit ourselves just to the treatment of such *continuous wave signals*.

The main mechanisms by which a NS can radiate gravitational waves consist of non-axisymmetric distortions in the solid part of the star (i.e. the case treated here, where the signal frequency f is twice the star rotation frequency f_r) [25, 26], free precession of the NS ($f = f_r$) [27] and fluid r-modes ($f \approx 4f_r/3$) [28, 29, 30].

4. The continuous wave signal

As already mentioned, continuous gravitational waves are expected to be produced by rapidly rotating NSs with non-axisymmetric deformations [30, 31]. The general form of a continuous gravitational wave signal is described by the following tensor metric perturbation:

$$\mathbf{h}(t) = h_+(t)\mathbf{e}_+(t) + h_\times(t)\mathbf{e}_\times(t), \tag{1}$$

where $h_+(t)$ and $h_\times(t)$ are the waveforms of the two orthogonal transverse polarizations, "+" and "×", respectively and are given by

$$h_+(t) = h_0\left(\frac{1+\cos^2\iota}{2}\right)\cos\Phi(t), \qquad h_\times(t) = h_0\cos\iota\sin\Phi(t), \tag{2}$$

with $\mathbf{e}_{+,\times}$ representing the two basis polarization tensors [32]; t is the time in the detector frame, ι is the inclination angle of the star's rotation axis with respect to the line of sight; $\Phi(t)$ is the signal phase function and h_0 is the amplitude expressed by

$$h_0 = \frac{4\pi^2 G}{c^4}\frac{I_{zz}\epsilon f^2}{d}. \tag{3}$$

The constant G is the gravitational constant; c represents the light speed; I_{zz} is the star's principal moment of inertia (assumed to be aligned with its spin axis), ϵ is the equatorial ellipticity of the star [33], d is the distance to the star and f represents the signal frequency. As the time-varying components of the mass quadrupole moment tensor are periodic with period half the star rotation period, the gravitational wave frequency f is twice the rotation frequency f_r.

The detector response to a metric perturbation is given by the known relation

$$h(t) = F_+(t,\alpha,\delta,\psi)h_+(t) + F_\times(t,\alpha,\delta,\psi)h_\times(t), \tag{4}$$

where α and δ are the source right ascension and declination, respectively, ψ is the polarization angle of the wave and $F_{+,\times}$ are the detector antenna pattern functions for the two orthogonal polarizations [33].

Assuming that all of the frequency's derivative, also denoted with the term of spin-down, is due to emission of gravitational radiation, we can relate ϵ to f_r and $\dot{f}_r$ [34]:

$$h_0^{\rm sd} = 8.06\times10^{-19} I_{38}\, d_{\rm kpc}^{-1}\sqrt{\frac{|(\dot{f}_r/{\rm Hz\,s^{-1}})|}{(f_r/{\rm Hz})}}, \tag{5}$$

Journal of Physics: Conference Series **354** (2012) 012010 doi:10.1088/1742-6596/354/1/012010

where I_{38} is the star's moment of inertia in units of the canonical value 10^{38} kg m^2 and $d_{\rm kpc}$ is the star's distance from the Sun in kiloparsec (kpc). This is referred to as *spin-down limit* on the signal amplitude and represents an absolute upper limit to the amplitude of the gravitational wave signal that could be emitted by the star, where electromagnetic radiation is neglected. The spin-down limit on strain corresponds to an upper limit on the star's ellipticity, given by [34]

$$\epsilon^{\rm sd} = 0.237 \left(\frac{h_0^{\rm sd}}{10^{-24}} \right) I_{38}^{-1} \, (f_r/{\rm Hz})^{-2} d_{\rm kpc}. \tag{6}$$

5. Data analysis methods and recent results

The way to search for continuous wave signals depends on how much about the source is known. There are different types of searches, briefly described in the text below and whose recent major results are also reported:

(i) *targeted searches*, where the source parameters (sky location, frequency, frequency derivatives) are assumed to be known with great accuracy (e.g. the Crab and Vela pulsars);

(ii) *directed searches*, where sky location is known while frequency and frequency derivatives are unknown (e.g. Cassiopeia A, SN1987A, Sco X-1, galactic center, globular clusters);

(iii) *all-sky searches* for unknown pulsars.

5.1. Targeted searches

This kind of searches is computationally cheap and a fully coherent analysis, based on matched filtering over long observation time, is quite feasible [33].

The minimum signal amplitude that can be detected over a given observation time $T_{\rm obs}$, assuming a certain false alarm probability (typically of 1 %) and a false dismissal probability (in general of 10 %) is given by

$$h_0^{\rm min} \approx 11 \sqrt{\frac{S_h(f)}{T_{\rm obs}}}. \tag{7}$$

Equation (7) is obtained by averaging over source and detector parameters and represents the sensitivity of a typical coherent search, with $S_h(f)$ being the detector noise power spectral density. Note that the precise value of the coefficient on the r.h.s of Eq. (7) depends on the analysis method employed (see for example Ref. [35] and references therein).

A search for continuous wave radiation from the Vela pulsar has been quite recently performed using data from the Virgo detector second science run (started on 2009 July 7 and ended on 2010 January 8) [34]. The resulting upper limits on continuous gravitational wave emission have been obtained using methods that assume the gravitational wave emission to follow the radio timing. Assuming known orientation of the star's spin axis and value of the wave polarization angle, frequentist upper limits of 1.9×10^{-24} and 2.2×10^{-24}, respectively, have been placed on the gravitational wave amplitude with 95 % confidence level. An independent method, under the same hypothesis, produces a Bayesian upper limit of 2.1×10^{-24} with 95 % degree of belief. These upper limits are well below the indirect spin-down limit of 3.3×10^{-24} for the Vela pulsar, defined by the energy loss rate inferred from observed decrease in Vela's spin frequency, and correspond to a limit on the star ellipticity of the order of 10^{-3}. Even assuming the star's spin axis inclination and the wave polarization angles unknown, the consequent results exhibit upper limits quite below the spin-down limit [34].

These recent results make Vela only the second pulsar for which the spin-down limit on gravitational wave emission has been beaten. The first pulsar for which this important result has been reached is the Crab pulsar [36, 37].

Another search worthy to be mentioned is the directed search for continuous wave signals from the non-pulsating NS in the supernova remnant Cassiopeia A over LIGO data. This search has established an upper limit on the signal amplitude over a wide range of frequencies which is below the indirect limit derived from energy conservation [38].

5.2. All-sky searches

It is well-known that all-sky searches for gravitational waves from unknown pulsars over wide-parameter spaces are computationally limited. The reason is that one needs to search for unknown sources located everywhere in the sky, with signal frequency as high as a few kHz and with values of spin-down as large as possible. Long integration times, typically of the order of a few months or years, are needed to build up sufficient signal power.

The data analysis strategy used to extract the faint continuous wave signals from the interferometric noise data was derived in Ref. [33] and is given by the standard coherent matched filtering method, that is based on the *maximum likelihood detection*. The resulting optimal coherent search statistic is the so-called $\mathcal{F}$-statistic.

Fully coherent methods based on matched filtering are the approach used in analyses for continuous wave searches over wide parameter space. However, they become computationally undoable when very long data stretches (of the order of months or years) are used and a wide fraction of the parameter space is searched over, because of the increasing number of templates [39]. Therefore, different incoherent hierarchical methods have been proposed [40, 35, 41]. In the hierarchical strategies, the entire data set is split into different shorter Fourier transformed data segments, which are then properly combined to account for Doppler shifts and spin-down. In other words, at first, every data chunk is analyzed coherently via matched filtering and afterward the information from the different segments is combined incoherently (that means that the phase information is lost). Three different methods have been developed that combine the results from the different segments incoherently, forming sums over power ("stack-slide" [40, 41] and "PowerFlux" [42, 43] schemes) or weighted binary counts ("Hough transform" [35, 44, 45]). The sums are then weighted according to the detector noise and antenna-pattern to maximize the signal-to-noise ratio.

The hierarchical methods are computationally faster than the standard coherent methods and have a comparable sensitivity.

In general, the whole data set, of duration $T_{\rm obs}$, is partitioned into N smaller segments of duration $T_{\rm coh}$ each. Given such N data segments, the typical sensitivity of a continuous wave all-sky search is given by

$$h_0^{\rm m} \approx \frac{25}{N^{1/4}}\sqrt{\frac{S_h(f)}{T_{\rm coh}}}, \tag{8}$$

where the exact numerical factor depends on the specific hierarchical method employed (see for instance Ref. [46] and references therein).

The output of a standard continuous wave hierarchical analysis is given by a set of candidates, i.e. points in the source parameter space with high values of a given statistic and which need a deeper study. Typically coincidences are done among the candidates obtained by the analysis over different data segments in order to reduce the false alarm probability [47]. The surviving candidates can be then analyzed coherently over longer time baselines in order to discard them or confirm detection.

Early LIGO data from S5 have been analyzed using two different methods. No gravitational waves could be claimed, but interesting upper limits have been placed. A first search used the first eight months of S5 [48], covering the full sky, the frequency band (50–1100) Hz and a range of spin-down values between -5×10^{-9} Hz/s and zero. At the highest frequency the search would have been sensitive to the gravitational radiation emitted by a NS placed at 500 pc with equatorial ellipticity larger than 10^{-6}. Another search was performed over the first two months of S5 using the Einstein@Home infrastructure [49, 47]. The analysis consisted of matched filtering over 30 hours-long data segments followed by incoherent combination of results via a concidence strategy. The analyzed parameter space consisted of the whole sky, the frequency interval (50 –1500) Hz and spin-down range between -2×10^{-9} Hz/s and zero. This search would have been sensitive to 90 % of signals in the frequency band (125–225) Hz with amplitude greater than 3×10^{-24}. The search sensitivity was estimated through Monte Carlo methods (injection of software simulated signals).

Three new Einstein@Home searches, covering the full two-years period of the LIGO S5 run, will be shortly published [50].

5.3. The Einstein@Home project

As already said, month-long coherent integration is necessary to accumulate a signal-to-noise ratio sufficient for detection. However, a powerful and effective method that allows us to use the longest possible coherent integration time, and thus improve the search sensitivity, is represented by distributing the computation through the volunteer computing project Einstein@Home [49]. Such a project is built upon the BOINC (Berkeley Open Infrastructure for Network Computing) architecture [51], namely a system that exploits the idle time on volunteer computers to solve scientific problems that require large amounts of computer power, such as to process data from gravitational wave detectors. At present, it provides roughly 300 TFlops of distributed computing resources.

A current Einstein@Home hierarchical search, analyzing data from S6 and using a new technique based on the $\mathcal{F}$-statistic global correlations [52], is expected to bring a relevant increase in terms of sensitivity. This is due to the longer coherent time baseline ($T_{\rm coh}$ = 60 hours) used in such a search.

6. Concluding remarks

Despite huge efforts on several fronts (i.e. the improvement of the detector sensitivities and the employment of efficient gravitational wave data analysis algorithms), to date no direct gravitational wave detection has been made, but relevant upper limits on gravitational wave signal strength have been derived. Moreover, with the advent of advanced LIGO and Virgo detectors [53, 54], an amazing improvement in strain sensitivity, of a factor ten with respect to their initial configuration, is expected to be reached in a few years after 2015. At this point, the era of gravitational wave astronomy will definitely begin and the possibility of a first direct gravitational wave detection will become much more concrete. The employment of robust and hard-hitting gravitational wave data analysis techniques will be crucial at that time [55, 56, 52].

Acknowledgments

The authors gratefully acknowledge the support of the United States National Science Foundation for the construction and operation of the LIGO Laboratory, the Science and Technology Facilities Council of the United Kingdom, the Max-Planck-Society, and the State of Niedersachsen/Germany for support of the construction and operation of the GEO600 detector, and the Italian Istituto Nazionale di Fisica Nucleare and the French Centre National de la Recherche Scientifique for the construction and operation of the Virgo detector. The authors also gratefully acknowledge the support of the research by these agencies and by the Australian Research Council, the International Science Linkages program of the Commonwealth of Australia, the Council of Scientific and Industrial Research of India, the Istituto Nazionale di Fisica Nucleare of Italy, the Spanish Ministerio de Educación y Ciencia, the Conselleria d'Economia Hisenda i Innovació of the Govern de les Illes Balears, the Foundation for Fundamental Research on Matter supported by the Netherlands Organisation for Scientific

Research, the Polish Ministry of Science and Higher Education, the FOCUS Programme of Foundation for Polish Science, the Royal Society, the Scottish Funding Council, the Scottish Universities Physics Alliance, The National Aeronautics and Space Administration, the Carnegie Trust, the Leverhulme Trust, the David and Lucile Packard Foundation, the Research Corporation, and the Alfred P. Sloan Foundation.

References

[1] Einstein A 1916 *Annalen der Physik* **49** 769
[2] Hulse R A and Taylor J H 1975 *Astrophys. J.* **195** 51.
[3] http://www.naic.edu/
[4] Leaci P 2008 *Development of Wideband Acoustic Gravitational Wave Detectors* (Unpublished doctoral dissertation) Univeristy of Trento (Italy)
[5] Abramovici A *et al* 1992 *Science* **256** 325
[6] Barish B and Weiss R 1999 *Phys. Today* **52** 44
[7] Bradaschia C *et al* 1990 *Nucl. Instum. Methods Phys. Res. A* **289** 518
[8] Weber J 1960 *Phys. Rev* **117** 306
[9] Baggio L. *et al* 2005 *Phys. Rev. Lett* **94** 241101
[10] Astone P *et al* 1997 *Astrop. Phys.* **7** 231
[11] Astone P *et al* 1993 *Phys. Rev. D* **47** 362
[12] Mauceli E *et al* 1996 *Phys. Rev. D* **54** 1264
[13] Bonaldi M *et al* 2006 *Phys. Rev. D* **74** 022003
[14] Leaci P *et al* 2007 *Phys. Rev. A* **76** 062101
[15] Leaci P *et al* 2008 *Phys. Rev. D* **77** 062001
[16] Leaci P *et al* 2008 *Class. Quantum Grav.* **25** 195018
[17] Danzmann K *et al* 1995 in *First Edoardo Amaldi Conference on Gravitational Wave Experiments* Ed. Coccia E, Pizzella G, Ronga F (World Scientific, Singapore)
[18] Tsubono K 1995 in *Gravitational Wave Experiments* Ed. Coccia E, Pizzella G, Ronga F (World Scientific, Singapore)
[19] Kuroda K 2006 *Class. Quantum Grav.* **23** S215
[20] Punturo M *et al* 2010 *Class. Quantum Grav.* **27** 194002
[21] http://lisa.nasa.gov/
[22] Kawamura k *et al* 2008 *J. Phys.: Conf. Ser.* **122** 012006
[23] Thorne K 1987 *300 Years of Gravitation* Hawking S W and Israel W, Cambridge University Press, Cambridge, England
[24] Thorne K 1995 *Gravitational waves in Proceedings of the Snowmass'94 Summer Study on Particle and Nuclear Astrophysics and Cosmology* eds. Kolb E W and Peccei R (World Scientific, Singapore)
[25] Cutler C 2002 *Phys. Rev. D* **66** 084025
[26] Ushomirsky G 2000 *Mon. Not. R. Astron. Soc.* **319** 902
[27] Stairs I *et al* 2000 *Nature* **406** 484
[28] Owen B J 1998 *Phys. Rev. D* **58** 084020
[29] Andersson N *et al* 1999 *Astrophys. J.* **516** 307
[30] Bildsten L 1998 *Astrophys. J.* **501** L89
[31] Owen B J 2005 *Phys. Rev. Lett.* **95** 211101
[32] Misner C, Thorne K, Wheeler J A 1973 *Gravitation* (San Francisco, CA: Freeman)
[33] Jaranowski P *et al* 1998 *Phys. Rev. D* **58** 063001
[34] Abadie J *et al* (LIGO and Virgo Scientific Collaboration) 2011 *Astrophys. J.* **737** 93
[35] Krishnan B *et al* 2004 *Phys. Rev. D* **70** 082001
[36] Abbott B *et al* (LIGO Scientific Collaboration) 2008 *Astrophys. J.* **683** L45
[37] Abbott B *et al* (LIGO and Virgo Scientific Collaboration) 2010 *Astrophys. J.* **713** 671
[38] Abadie J *et al* (LIGO Scientific Collaboration) 2010 *Astrophys. J.* **722** 1504
[39] Jaranowski P and Królak A 2000 *Phys. Rev. D* **61** 062001
[40] Brady P R and Creighton T 2000 *Phys. Rev. D* **61** 082001
[41] Cutler C *et al* 2005 *Phys. Rev. D* **72** 042004
[42] Abbott B *et al* (LIGO Scientific Collaboration) 2008 *Phys. Rev. D* **77** 022001
[43] Abbott B *et al* (LIGO Scientific Collaboration) 2009 *Phys. Rev. Lett.* **102** 111102
[44] Abbott B *et al* (LIGO Scientific Collaboration) 2005 *Phys. Rev. Lett.* **72** 102004
[45] Krishnan B (for the LIGO Scientific Collaboration) 2005 *Class. Quantum Grav* **22** S1265
[46] Abbott B *et al* (LIGO Scientific Collaboration) 2005 *Phys. Rev. D* **72** 102004

[47] Abbott B *et al* (LIGO Scientific Collaboration) 2009 *Phys. Rev. Lett.* **80** 042003
[48] Abbott B *et al* (LIGO Scientific Collaboration) 2009 *Phys. Rev. Lett.* **102** 11102
[49] http://einstein.phys.uwm.edu
[50] Abadie J *et al* (LIGO and Virgo Scientific Collaboration) *New search for periodic gravitational waves in LIGO S5 data through Einstein@Home grid computing system* (in preparation)
[51] http://boinc.berkeley.edu/
[52] Pletsch H J 2008 *Phys. Rev. D* **78** 102005
[53] http://www.ligo.caltech.edu/advLIGO/scripts/summary.shtml
[54] https://wwwcascina.virgo.infn.it/
[55] Leaci P *et al* 2010 *J. Phys.: Conf. Ser.* **228** 012006
[56] Leaci P *et al* *New methods to veto spurious candidates in continuous wave searches from gravitational wave detectors* (in preparation)

Journal of Physics: Conference Series **354** (2012) 012011 doi:10.1088/1742-6596/354/1/012011

Fermionic DBI and Chaplygin gas unified models of dark energy and dark matter from f-essence

D Momeni
Eurasian International Center for Theoretical Physics, Eurasian National University, Astana 010008, Kazakhstan
E-mail: d.momeni@yahoo.com

Y Myrzakulov
Eurasian International Center for Theoretical Physics, Eurasian National University, Astana 010008, Kazakhstan

P Tsyba
Eurasian International Center for Theoretical Physics, Eurasian National University, Astana 010008, Kazakhstan

K Yesmakhanova
Eurasian International Center for Theoretical Physics, Eurasian National University, Astana 010008, Kazakhstan

R Myrzakulov
Eurasian International Center for Theoretical Physics, Eurasian National University, Astana 010008, Kazakhstan
E-mail: rmyrzakulov@csufresno.edu; rmyrzakulov@gmail.com

Abstract. In the present work, we study the cosmological model with fermionic field and with the non-canonical kinetic term (*f-essence*). We present some important reductions of the model as well as some of its generalizations. We also find the exact solution of the model and examine the influence of such gravity-fermion interaction on the observed accelerated expansion of our universe. Further, we discuss the cosmological implications of the exact models. We show that our f-essence model corresponds to ΛCDM, but not a standard cold dark matter model containing no radiation. Also, our particular f-essence model has no Einstein's static universe. Further we discuss several fermionic Dirac-Born-Infeld (DBI) models as some particular reductions of f-essence, respectively and their Chaplygin gas counterparts are found.

1. Introduction

The recent data from type Ia Supernovae (SNIa) and cosmic microwave background (CMB) radiation have provided strong evidence for a spatially flat and accelerated expanding universe [1]-[2].

In the context of Friedmann-Robertson-Walker (FRW) cosmology with Einstein gravity, this acceleration is attributed to the domination of a component with negative pressure, called dark energy. So far, the nature of dark energy remains a mystery. Theoretically, the simplest candidate for such a component is a small positive cosmological constant, but it suffers the difficulties associated with the fine tuning and the coincidence problems [3]. Therefore, many authors are attracted by the idea of dynamical dark energy models, such as quintessence, phantom, k-essence, quintom etc (see Refs. [4]-[6] for recent reviews).

Assuming the validity of the theory of gravity, one attempt to explain the present stage of accelerated expansion of our universe is the existence of an exotic, currently dominant, ingredient of the energy content of the universe, known as dark energy, with unusual physical properties. The other possibility is modifying the general theory of relativity at large scales [7, 8, 9]. In cosmology, the investigation for the constituents responsible for the accelerated periods in the evolution of the universe is of great interest. The mysterious dark energy has been proposed as a cause for the late time dynamics of the current accelerated phase of the universe.

Recently, theories described by an action with non-standard kinetic terms, k-essence, attracted considerable interest. Such theories were first studied in the context of k-inflation [10], and then the k-essence models were suggested as dynamical dark energy for solving the cosmic coincidence problem [11]-[14]. The k-essence models have also attractor solutions: it means that the differential equations of k-essence model give solution representing the present state of the universe regardless of the choice of initial conditions imposed on the value of the scalar field. Thus the coincidence problem is resolved in k-essence model. Such models of k-essence with non canonical forms of the Lagrangian in some cases, cannot define a definite dynamics for the scalar field. Recently, several attempts were made to explain the accelerated expansion by choosing fermionic fields as the gravitational sources of energy (see e.g. refs. [15]-[33]). In particular, it was shown that the fermionic field plays a very important role in: i) keeping the isotropic shape of an initially anisotropic spacetime; ii) cosmological solutions without singularity ; iii) prediction of late-time acceleration. In the present work, we study the cosmological model with fermionic field — the so-called f-essence. We examine the influence of such gravity-fermionic interactions on the accelerated expansion of the Universe. The formulation of the gravity-fermionic theory has been discussed in detail elsewhere [34]-[37], so we will only present the result here.

2. Basics of f-essence

In this work we consider the f-essence fermionic field ψ minimally coupled to the gravitational field $g_{\mu\nu}$ given by the action

$$S = \int d^4x\sqrt{-g}[R + 2K(Y, \psi, \bar{\psi})]. \tag{1}$$

Here the Lagrangian K depends on the fermionic field ψ, its conjigate $\bar{\psi}$ and the canonical kinetic term Y. In general, the canonical kinetic term Y reads as

$$Y = 0.5i[\bar{\psi}\Gamma^{\mu}D_{\mu}\psi - (D_{\mu}\bar{\psi})\Gamma^{\mu}\psi]. \tag{2}$$

We work with a space-time metric of the form

$$ds^2 = -dt^2 + a^2(dx^2 + dy^2 + dz^2), \tag{3}$$

that is the FRW metric. For this metric, the vierbein is chosen to be

$$(e_a^{\mu}) = diag(1, 1/a, 1/a, 1/a), \quad (e_{\mu}^a) = diag(1, a, a, a). \tag{4}$$

The Dirac matrices of curved spacetime Γ^{μ} are

$$\Gamma^0 = \gamma^0, \quad \Gamma^1 = a^{-1}\gamma^1, \quad \Gamma^2 = a^{-1}\gamma^2, \quad \Gamma^3 = a^{-1}\gamma^3, \tag{5}$$

$$\Gamma_0 = \gamma^0, \quad \Gamma_1 = a\gamma^1, \quad \Gamma_2 = a\gamma_2, \quad \Gamma_3 = a\gamma_3. \tag{6}$$

Hence we get

$$\Omega_0 = 0, \quad \Omega_1 = 0.5\dot{a}\gamma^1\gamma^0, \quad \Omega_2 = 0.5\dot{a}\gamma^2\gamma^0, \quad \Omega_3 = 0.5\dot{a}\gamma^3\gamma^0. \tag{7}$$

We now ready to write the equations of the f-essence. We have

$$3H^2 - \rho = 0, \tag{8}$$
$$2\dot{H} + 3H^2 + p = 0, \tag{9}$$
$$K_Y\dot{\psi} + 0.5(3HK_Y + \dot{K}_Y)\psi - i\gamma^0 K_{\bar{\psi}} = 0, \tag{10}$$
$$K_Y\dot{\bar{\psi}} + 0.5(3HK_Y + \dot{K}_Y)\bar{\psi} + iK_{\psi}\gamma^0 = 0, \tag{11}$$
$$\dot{\rho}_f + 3H(\rho_f + p_f) = 0, \tag{12}$$

where $Y = 0.5i(\bar{\psi}\gamma^0\dot{\psi} - \dot{\bar{\psi}}\gamma^0\psi)$ and

$$\rho_f = YK_Y - K, \quad p_f = K. \tag{13}$$

are the energy density and pressure of the fermionic field. If $K = Y - V$, then from the system (8)-(12) we get the corresponding equations of the Einstein-Dirac model.

3. Solution

In this subsection we want to construct a solution of the particular f-essence model. Let $K = K(Y, u)$, where $u = \bar{\psi}\psi$.
We now consider the submodel, where we assume that $A_2 = \alpha Y^n$, $B_2 = \beta u^m$ that is the case $K = \alpha Y^n + \beta u^m$. Let $a = a_0 t^{\lambda}$. Then we have the following solution

$$Y = \left\{\left[-\frac{6m\lambda^2 - 4(m+1)\lambda}{\alpha m}\right]t^{-2}\right\}^{\frac{1}{n}}, \quad u = \left\{\left[\frac{-4\lambda}{\beta m}\right]t^{-2}\right\}^{\frac{1}{m}}, \tag{14}$$

where

$$\lambda = \frac{2n - 2m + 2mn}{3mn}, \quad a_0 = \sqrt[3]{\frac{c}{\alpha n}\left[-\frac{6m\lambda^2 - 4(m+1)\lambda}{\alpha m}\right]^{\frac{1-n}{n}}\left[-\frac{4\lambda}{\beta m}\right]^{-\frac{1}{m}}}. \tag{15}$$

Finally we present the following formulas

$$u = ca^{-3}K_Y^{-1}, \quad \psi_j = c_j a^{-1.5}K_Y^{-0.5}e^{i\gamma^0\int K_2' K_Y^{-1}dt}, \tag{16}$$

where $c = |c|_1^2 + |c|_2^2 - |c|_3^2 - |c|_4^2$, $c_j = consts.$

3.1. *The cosmological implications of the f-essence model*

Our model for some values of the parameters (m, n) may be accelerating or decelerating. But in favor of the observational data we must have accelerating expansion today. The necessary and sufficient condition to have acceleration expansion is the $\ddot{a} > 0$. From $a = a_0 t^{\lambda}$ we observe that for an acceleration phase we must have $\lambda(\lambda - 1) > 0$. It means that we must have

$\lambda < 0$, *or*, $\lambda > 1$. By assuming that $a_0 > 0$, we must check the following function of the two variables (m, n) for positivity

$$f(m,n) \equiv \lambda(\lambda - 1) = -\frac{2}{9}\frac{(n - m + mn)(-2n + 2m + mn)}{m^2 n^2} \tag{17}$$

Hence we find just two values of λ give us the accelerated expansion.
The declaration parameter q in terms of the Hubble parameter is defined by

$$q = -(1 + \frac{\dot{H}}{H^2}) \tag{18}$$

Since $H = \frac{\lambda}{t}$, then it reads

$$q = -(1 - \frac{1}{\lambda}) \tag{19}$$

To have an expanding universe , $q > -1$. Thus we must have $\lambda > 0$. This condition in comparison to (17) indicates that the only possible value of the parameter λ to having both acceleration and expansion is $\lambda > 1$. In terms of the set of the parameters (m, n) it results

$$\frac{2n - 2m + 2mn}{3mn} > 1 \tag{20}$$

indeed for (20), there are four different kinds of the solutions which we summarize them here:
Case 1: $m, n > 0$. In this case, we have the following acceptable solutions

$$m > \frac{2n}{n+2}. \tag{21}$$

This is a condition which under it, the model describes the acceleration expansion.

Case 2: $m, n < 0$. In this case, we have two possibility

$$m > \frac{2n}{n+2}, \quad n + 2 > 0. \tag{22}$$

$$m < \frac{2n}{n+2}, \quad n + 2 < 0. \tag{23}$$

With both of these conditions, we conclude that the model describes the acceleration expansion.

Case 3: $m > 0, n < 0$. As the case 2, in this case, we have two possibility

$$m < \frac{2n}{n+2}, \quad n + 2 > 0. \tag{24}$$

$$m > \frac{2n}{n+2}, \quad n + 2 < 0. \tag{25}$$

Both of these conditions indicate that the model describes the acceleration expansion.

Case 4: $m < 0, n > 0$. As the case 1, in this case, there is only one possibility

$$m < \frac{2n}{n+2}. \tag{26}$$

Thus our solution for some values of the parameters m, n describes the acceleration expansion.

Now we discuss the statefinder parameters $\{r, s\}$. They have been introduced to characterize the spatially flat universe $k = 0$ models with cold dark matter (dust) and dark energy [48]. They were defined by the following expressions:

$$r \equiv \frac{\ddot{a}}{aH^3} = \frac{t(\lambda - 1)}{\lambda^2}, \tag{27}$$

$$s \equiv \frac{r-1}{3(q-\frac{1}{2})} = -\frac{2}{3}\frac{t\lambda - t - \lambda^2}{\lambda\,(3\,\lambda - 2)}. \tag{28}$$

We obtain the relation between r and s as

$$s = -\frac{2}{3}\frac{(r-1)\,\lambda}{3\lambda - 2}. \tag{29}$$

The Fig. 1 shows the plot of the (29) for some values of the $\lambda > 1$.

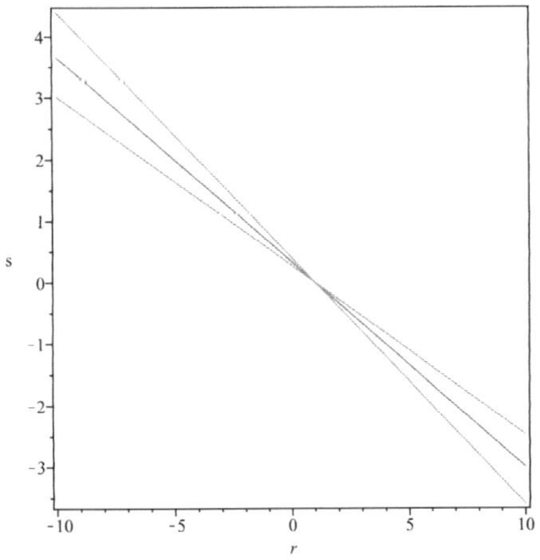

Figure 1. Plot of the (29) for some values of the $\lambda > 1$

The cosmological constant possesses a fixed equation of state parameter ($w = -1$) and a fixed Newton's gravitational constant. Here $\{1, 0\}$ corresponds to ΛCDM. Moreover $\{1, 1\}$ represents the standard cold dark matter model containing no radiation while Einstein's static universe corresponds to $\{-\infty, \infty\}$ [49]. It is interesting to note that from (29) we can conclude that our f-essence model corresponds to ΛCDM, since it possesses the point $\{1, 0\}$ for any value of $\lambda \neq \frac{2}{3}$. The point $\{1, 1\}$ which represents the standard cold dark matter model containing no radiation is absent in our f-essence model. Also, our f-essence model has no analogue of the Einstein static universe corresponding to the fixed point $\{-\infty, \infty\}$.

4. f-DBI

Let us return to the the action (1)

$$S = \int d^4x\sqrt{-g}[R + 2K(Y, \psi, \bar{\psi})]. \tag{30}$$

For the FRW metric the corresponding system of equations takes the form (8)-(12). Now we want to present some examples of fermionic DBI models (or f-DBI for short. First it is necessary to review some essentials of the DBI and tachyonic models. In the past, the use of the tachyon as a scalar field with negative and non canonical kinetic energy has been explored in certain string theories. Tachyons are used both for cosmology and specially for better understanding of the D- brane decay process [50]. This led to study of the role of tachyons in cosmology as a rolling tachyon, which is a scalar field ϕ has an EoS parameter between $(-1, 0)$ [51]. Thus a tachyon may be a suitable candidate for inflation in the high energy regime [52]. It can also be regarded as a candidate for dark energy even in some extended versions of Einstein gravity [53].

4.1. Tachyonic models

Let us we present some examples of tachyonic models that follow from f-essence. Consider examples.
i) First let us consider the following tachyonic models

$$K = -U\sqrt{1-\alpha Y}, \tag{31}$$

where

$$U = U(\bar{T}, T), \quad Y = 0.5i(\bar{T}\gamma^0\dot{T} - \dot{\bar{T}}\gamma^0 T) \tag{32}$$

and T is the fermionic tachyon field. Then we get the following EoS

$$p = -\rho \mp \sqrt{\rho^2 - U^2}, \tag{33}$$

where

$$\rho = \frac{0.5U}{\sqrt{1-\alpha Y}} + 0.5U\sqrt{1-\alpha Y}. \tag{34}$$

ii) Let the f-essence Lagrangian has the form

$$K_2 = -U\sqrt{1-\alpha Y^2}, \tag{35}$$

where

$$U = U(\bar{T}, T), \quad Y = 0.5i(\bar{T}\gamma^0\dot{T} - \dot{\bar{T}}\gamma^0 T) \tag{36}$$

and T is the fermionic tachyon field. Then we get the following EoS

$$p = -\frac{U^2}{\rho}, \tag{37}$$

where

$$\rho = \frac{U}{\sqrt{1-\alpha Y^2}}. \tag{38}$$

So this model corresponds to the Chaplygin gas [38].

4.2. Dark energy models

a) Now we consider the following DBI model

$$K_2 = \epsilon U(\sqrt{1-\alpha\frac{Y}{U}} - 1) + V, \tag{39}$$

where

$$U = U(\bar{\psi}, \psi), \quad Y = 0.5i(\bar{\psi}\gamma^0\dot{\psi} - \dot{\bar{\psi}}\gamma^0\psi), \quad V = V(\bar{\psi}, \psi) \tag{40}$$

and ψ is the classical commuting fermionic field. Then we get the following EoS

$$p = -2\bar{\rho} \pm \sqrt{4\bar{\rho}^2 - 2\epsilon^2 U^2} + V - \epsilon U, \tag{41}$$

where $\bar{\rho} = \rho - \epsilon U + V$ and the energy density is given by

$$\rho = -0.5\epsilon U\sqrt{1-\alpha\frac{Y}{U}} - \frac{0.5\epsilon U}{\sqrt{1-\alpha\frac{Y}{U}}} + \epsilon U - V. \tag{42}$$

b) Our second example reads as

$$K = \epsilon U(\sqrt{1-\alpha\frac{Y^2}{U}} - 1) + V, \tag{43}$$

where

$$U = U(\bar{\psi}, \psi), \quad Y = 0.5i(\bar{\psi}\gamma^0\dot{\psi} - \dot{\bar{\psi}}\gamma^0\psi), \quad V = V(\bar{\psi}, \psi) \tag{44}$$

and ψ is the classical commuting fermionic field. Then we get the following EoS

$$p = -\frac{\epsilon^2 U^2}{\rho - \epsilon U + V} - \epsilon U + V, \tag{45}$$

where the energy density is given by

$$\rho = -\frac{\epsilon U}{\sqrt{1-\alpha\frac{Y^2}{U}}} + \epsilon U - V. \tag{46}$$

As $V = \epsilon U$ this model becomes

$$p = -\frac{\epsilon^2 U^2}{\rho} \tag{47}$$

that corresponds to the Chaplygin gas [38]. We noted here that the correspondence between DBI essence and Chaplygin gas models has been discussed previously [54].

5. Integrable f-essence models

One of interesting class of f-essence models is integrable models (see e.g. [39]-[42]). Let the Lagrangian of f-essence has the form

$$K_2 = F(Y) - V(\bar{\psi}, \psi). \tag{48}$$

As an example, here we consider the following particular f-essence model when $F(Y)$ obeys the equation

$$F_{YY} = 2F^3 + YF + \alpha, \tag{49}$$

where α is some constant. It is nothing but the P_{II} - equation so that it is integrable. Now the full equations of f-essence take the form

$$3H^2 - \rho_f = 0, \tag{50}$$

$$2\dot{H} + 3H^2 + p_f = 0, \tag{51}$$

$$F_Y\dot{\psi} + 0.5(3HF_Y + \dot{F}_Y)\psi + i\gamma^0 V_{\bar{\psi}} = 0, \tag{52}$$

$$F_Y\dot{\bar{\psi}} + 0.5(3HF_Y + \dot{F}_Y)\bar{\psi} - iV_{\psi}\gamma^0 = 0, \tag{53}$$

$$F_{YY} - 2F^3 - YF - \alpha = 0, \tag{54}$$

$$\dot{\rho}_f + 3H(\rho_f + p_f) = 0. \tag{55}$$

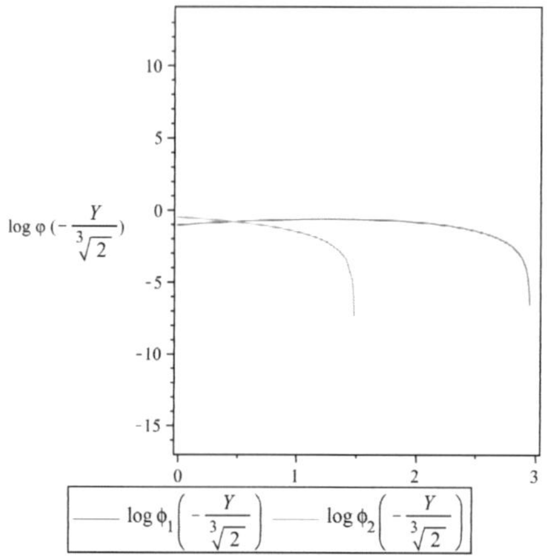

Figure 2. The plot of the unnormalized $\log(\phi(-2^{-1/3}Y))$

To construct solutions of this system we start from the equation (54). The equation (54) has the following particular solutions (see e.g. [39]-[47])

$$F \equiv F(Y;1.5) = \psi - (2\psi^2 + Y)^{-1}, \tag{56}$$

$$F \equiv F(Y;1) = -\frac{1}{Y}, \tag{57}$$

$$F \equiv F(Y;2) = \frac{1}{Y} - \frac{3Y^2}{Y^3+4}, \tag{58}$$

$$F \equiv F(Y;3) = \frac{3Y^2}{Y^3+4} - \frac{6Y^2(Y^3+10)}{Y^6+20Y^3-80}, \tag{59}$$

$$F \equiv F(Y;4) = -\frac{1}{Y} + \frac{6Y^2(Y^3+10)}{Y^6+20Y^3-80} - \frac{9Y^5(Y^3+40)}{Y^9+60Y^6+11200}, \tag{60}$$

$$F \equiv F(Y;0.5\epsilon) = -\epsilon\psi \tag{61}$$

and so on. Here

$$\psi = (\ln\phi)_Y, \quad \phi(Y) = C_1 Ai(-2^{-1/3}Y) + C_2 Bi(-2^{-1/3}Y), \tag{62}$$

where $C_i = consts$ and $Ai(x), Bi(x)$ are Airy functions.

The Fig. 2 shows the plot of the unnormalized $\log(\phi(-2^{-1/3}Y))$ for $C_1 = 1, C_2 = 0$(we denote it by $\log(\phi_1(-2^{-1/3}Y))$) and for $C_1 = 0, C_2 = 1$(we denote it by $\log(\phi_2(-2^{-1/3}Y))$).

6. Conclusion

The accelerating expansion of the universe is one of the biggest challenges in modern physics. If we keep the Einstein theory as the geometrical part of the model and think of the matter fields as the sources of such an acceleration, we have many choices for it. One of them is scalar field, with canonical or non canonical form. The canonical scalar field can not describe the acceleration expansion correctly and it is usefull only in inflationary scenarios. A natural extension of the canonical scalar field, is a non canonical ones in which we are working with a general form of the kinetic energy. There are diffrent kinds of such models as k-essence. In this paper we considered a new model, f-essence in which the classical Dirac fermions play the role of the source of the dark energy. We studied the particular Lagrangian of such models. In the case of no interaction between the scalar and fermion fields (no Yukawa-like interaction) we obtained some exact solutions. For this kind of solution we show that the model has an accelerating expansion and the scale factor behaves power law. By analyzing the statefinder parameters $\{r, s\}$, we showed that our f-essence model corresponds to ΛCDM, but not a standard cold dark matter model containing no radiation. Also, our exact f-essence model can not has any Einstein's static universe. Further we consider several fermionic DBI models as some particular reductions of f-essence and their Chaplygin gas counterparts are found. We showed that we can reconstruct the DBI and tachyonic and Chaplygin gas models in such a f-essence model with some specific forms of the Lagrangian.

7. References

[1] Perlmutter S. et al.1999 Astrophys. J. **517**, 565-586
[2] Riess A G 1998 Astron.J.116,1009
[3] Peebles P J E, Ratra, B 2003 Rev. Mod. Phys. 75, 559
[4] Copeland E J, Sami M, Tsujikawa S 2006 Int. J. Mod. Phys. D, **15**, 1753
[5] Frieman J, Turner M, Huterer D 2008 Ann. Rev. Astron. Astrophys. **48**, 385
[6] Caldwell R R, Kamionkowski M 2009 Ann. Rev. Nucl. Part. Sci. **59**, 397
[7] Nojiri S, Odintsov S D 2007 Int.J.Geom.Meth.Mod.Phys.4,115;
Montelongo Garci'a N , Harko T, Lobo F. S. N, Mimoso J P 2011 J.Phys.Conf.Ser.314:012060 ;
Francisco T, Lobo S N , Nojiri S, Odintsov S D 2011 Phys.Rev.D84:024020;
Raza M ,Jamil M ,Momeni D ,Myrzakulov R 2011 (Preprint arXiv:1107.5807);
Myrzakulov R 2011 Eur.Phys.J.C71:1752)
[8] Nojiri S, Odintsov S D 2011 Phys. Rept. **500**, 59-144
[9] Tsujikawa S 2010 Lect. Notes Phys. **800**, 99-145
[10] Armendariz-Picon C, Damour T, Mukhanov V F 1999 Phys. Lett. **B458**, 209-218
[11] Armendariz-Picon C, Mukhanov V F , Steinhardt P J 2001 Phys. Rev. **D63**, 103510
[12] Armendariz-Picon C, Mukhanov V F , Steinhardt P J 2000 Phys. Rev. Lett. **85**, 4438-4441
[13] Chiba T, Okabe T, Yamaguchi M 2000 Phys. Rev. **D62**, 023511
[14] Tsyba P Y, Kulnazarov I I, Yerzhanov K K, Myrzakulov T 2011 International Journal of Theoretical Physics, **50**, 1876-1886
[15] Ribas M O., Devecchi F P , Kremer G M 2005 Phys. Rev.**D72**, 123502
[16] Samojeden L L, Devecchi F P, Kremer G M 2010 Phys. Rev. **D81**, 027301
[17] Samojeden L L, Kremer G M, Devecchi F P 2009 Europhys. Lett. **87**, 10001
[18] Ribas M O, Kremer G M 2010 Grav. Cosmol. **16**, 173-177
[19] Cai Y F, Wang J 2008 Class. Quant. Grav., **25**, 165014
[20] Wang J, Cui S-W., Zhang C-M 2010 Phys. Lett., **B683**, 101-107
[21] Ribas M O, Devecchi F P, Kremer G M 2008 Europhys. Lett. **81**, 19001
[22] Rakhi R, Vijayagovindan G V, Indulekha K 2009 *A cosmological model with fermionic field*,[Preprint arXiv:0912.1222]
[23] Rakhi R, Vijayagovindan G V, Noble P A, Indulekha K 2010 International Journal of Modern Physics A, **25**, 1267-1278
[24] Chimento L P, Devecchi F P, Forte M , Kremer G M 2008 Class. Quant. Grav. **25**, 085007
[25] Anischenko S V, Cherkas S L, Kalashnikov V L 2004 *Cosmological Production of Fermions in a Flat Friedmann Universe with Linearly Growing Scale Factor: Exactly Solvable Model* [Preprint arXiv:0911:0769]
[26] Saha B 2004 Phys. Rev. **D69**, 124006
[27] Esmakhanova K, Myrzakulov Y, Nugmanova G, Myrzakulov R 2011 International Journal of Theoretical Physics, [DOI: 10.1007/s10773-011-0996-3]
[28] Razina O V, Myrzakulov Y M, Serikbayev N S, Myrzakul Sh R, Nugmanova G N, Myrzakulov R 2011 European Physical Journal Plus, **126**, N11, 1698
[29] Razina O V, Myrzakulov Y M, Serikbayev N S, Nugmanova G N, Myrzakulov R 2011 Central European Journal of Physics, [DOI:10.2478/s11534-011-0102-8]
[30] Vakili B, Sepangi H R 2008 Annals Phys. **323**, 548-565
[31] Dereli T, Ozdemir N, Sert O 2010 *Einstein-Cartan-Dirac Theory in (1+2)-Dimensions*, [Preprint arXiv:1002.0958]
[32] Balantekin A B, Dereli T 2007 Phys. Rev. **D75**, 024039
[33] Armendariz-Picon C, Greene P 2003 Gen Relativ Gravit **35**, 1637-1658

[34] Weinberg S 2007 *Gravitation and Cosmology* (John Wiley & Sons, New York, 1972), ibid. *Cosmology* (Cambridge, New York)
[35] Wald R M 1984*General Relativity*, (The University of Chicago Press, Chicago,
[36] Ryder L H 1996 *Quantum Field Theory* (Cambridge University Press, Cambridge)
[37] Birrell N D, Davies P C W 1982*Quantum Fields in Curved Space* (Cambridge University Press, Cambridge)
[38] Kamenshchik A Y, Moschella U, Pasquier V 2001 Phys. Lett. B, **511**, 265
[39] Esmakhanova K, Myrzakulov N, Nugmanova G, Myrzakulov Y, Chechin L, Myrzakulov R 2011 *Dark energy in some integrable and nonintegrable FRW cosmological models*, International Journal of Modern Physics D, **20**, N12, 1-28 [DOI No: 10.1142/S0218271811020445].
[40] Nugmanova G N, Myrzakul Sh R, Razina O V, Esmakhanova K R, Serikbayev N S, Myrzakulov R 2011 *Some cosmological aspects of Horava-Lifshitz gravity: integrable and nonintegrable models* [Preprint arXiv:1104.5374]
[41] Myrzakulov R 2011 Eur. Phys. J. C, **71**, 1752
[42] Myrzakul Sh, Esmakhanova K, Myrzakulov K, Nugmanova G, Myrzakulov R 2011 *FRW cosmological models with integrable and nonintegrable differential equations of state*, [Preprint arXiv:1105.2771]
[43] Kulnazarov I, Yerzhanov K, Razina O, Myrzakul Sh, Tsyba P, Myrzakulov R 2011 Eur. Phys. J. C, **71**, 1698
[44] Dzhunushaliev V, Folomeev V, Myrzakulov R 2010 Phys. Lett. B, **693**, 209-212
[45] Elizalde E, Myrzakulov R, Obukhov V V, Saez-Gomez D 2010 Classical and Quantum Gravity, **27**, 095007
[46] Myrzakulov R, Saez-Gomez D, Tureanu A 2011 General Relativity and Gravitation, **43**, 1671-1684
[47] Jamil M, Myrzakulov Y, Razina O, Myrzakulov R 2011 Astrophys. Space Sci., **336**, N2, 315-325
[48] Sahni V et al 2003 JETP Lett. 77 201
[49] Jamil M , Debnath U 2011 International Journal of Theoretical Physics 50:1602-1613[Preprint arXiv:0909.3689]
[50] Sen A 2003 International Journal of Modern Physics A 18 4869;
Gorini V, et al 2004 Phys. Rev. D 69 123512
[51] Gibbons G W 2002 Phys Lett B 537 1
[52] Mazumdar A, Panda S, Perez-Lorenzana A 2001, Nucl. Phys. B 614 101
[53] Jamil M et al 2011 Phys Lett B 702 315-319
[54] Debnath U , Jamil M 2011 Astrophys. Space Sci. 335:545-552.

Journal of Physics: Conference Series **354** (2012) 012012 doi:10.1088/1742-6596/354/1/012012

Effect of Accretion of Phantom Energy on Initial Mass of a Primordial Black Hole

Sumaira Naz and Asghar Qadir

Center for Advanced Mathematics and Physics,
National University of Sciences and Technology (NUST), H-12, Islamabad, Pakistan

E-mail: math_uniqueness@yahoo.com, aqadirmath@yahoo.com

Abstract. Hawking had shown that black holes radiate with a temperature inversely proportional to their mass, thereby losing energy and hence mass. For sufficiently small masses (less than 10^{15} g) the black hole would evaporate today and hence has a "life" equal to the present age of the universe. One explanation of the observed acceleration of the universe is by phantom energy. In 2010 Jamil and Qadir have showed that this energy enhances the rate of evaporation. Thus, to have a primordial black hole evaporating today, its initial mass should be larger than 10^{15} g or the primordial black holes of mass less than 10^{15} g should evaporate earlier. In fact, it was claimed that the black holes would be ten orders of magnitude larger! This effect is revisited and its dependence on the value of equation of state parameter is studied. It is found that the effect of phantom energy for the 10^{15} g black hole is negligible but for a 10^{22} g black hole would be significant. In that case, though, the black hole would not be now exploding. The mass at which the effect of phantom energy equals the effect of Hawking radiation has been called the transition mass. The transition mass has been discussed and the correction term in the lifetime for primordial black holes is computed.

1. Introduction

In an attempt to construct a theory of quantum gravity, as a "half-way-house", Hawking tried quantizing scalar fields in a classical curved space-time background of a Schwarzschild black hole. He found that there would be radiation at a temperature inversely proportional to the mass [1]. Consequently the black hole loses mass and therefore the temperature rises. The more it loses the faster is the rise, leading to an explosion burst of radiation in which the black hole loses its mass totally. As such sufficiently small mass Primordial Black Holes (the black holes thought to be formed within the first 100s after the big bang [2], PBHs hereafter) could evaporate today [1]. If PBHs are to evaporate today due to Hawking radiation (HR hereafter) only, then their initial mass would have been less than 10^{15} g and they must have formed within the first $\sim 10^{-23}$ s [3]. Although there is no evidence of their formation, their study is of great interest because their formation could provide a unique probe of the early Universe (if $M_{PBH} < 10^{15}$ g), gravitational collapse (if $M_{PBH} > 10^{15}$ g), high energy physics (if $M_{PBH} \sim 10^{15}$ g), quantum gravity (if $M_{PBH} < 10^{-5}$ g) and may place constraints on the physics relevant to these areas even if they never existed [3].

The universe is generally believed to have an accelerated expansion [4] despite claims that the observations could be accounted for by the inhomogeneities in the universe [5, 6]. The accelerated expansion is consistent with a cosmological constant but is generally taken to be due to some field that arises some time after the Plank era [7]. This field has been variously taken to be quintessence, dark energy [8] (DE hereafter) or phantom energy [9]. These exotic forms of energy have an equation of state (EoS) $\omega \equiv p/\rho$. For phantom energy $\omega < -1$, hence violating the dominant energy condition, a notion that helps prohibit time machine and wormholes [9].

Phantom energy is found to be compatible with most classical tests of cosmology based on current data as well as cosmic microwave background (CMB) anisotropy and mass power spectrum. The energy density of phantom energy increases with time [9]. It becomes infinite in finite time overcoming all other forms of matter and ripping apart the fabric of the universe [10]. It was shown that the accretion of phantom energy on a black hole effectively decreases its mass [11, 12, 13, 14]. In particular, Jamil and Qadir [13] have discussed the decrease in mass of a PBH due to the combined effect of accretion of phantom energy and Hawking evaporation. A relation between the time and mass of a PBH (which evaporates at $t = t_f$) was derived and it was said that the accretion of phantom energy on a black hole decreases its life or in other words we have to increase its initial mass to keep the lifetime of a black hole (BH hereafter) the same. The evolution of PBHs under the combined effect of HR, accretion of phantom energy and radiation accretion is studied by Guariento et al in 2008 [14]. In 2011 Nayak and Singh have discussed the increase in lifetime of PBHs when radiation accretion is taken into account in addition to HR [15]. Also in 2011, Nayak and Jamil have discussed the accretion of matter, radiation and vacuum energy on PBHs [16].

The plan of the paper is as follows. In section 2, we have repeated the calculations of [13] to correct a few mistakes and found that the results (formula, graphs and hence their interpretation) are different; increasing the ratio of the phantom energy mass reduction coefficient to that of the HR does not lead to any odd features. In the next section we calculate the transition masses of PBHs, i.e., the values of mass below which HR becomes more effective than phantom energy [14]. This depends on the time (t_o) when DE starts dominating over matter. Since at present the dark energy is more than twice the matter energy, it is reasonable to take t_o to be somewhere between 1 and 10 billion years. We have calculated the transition masses at the two extreme ends of this range. Taking into account the increased evaporation rate of PBHs, we compute the decrease in the "life" of PBHs of mass 10^{15} g in section 4 and find that this decrease is insignificant. However the effect is significant for larger mass BHs ($\sim 10^{22}$ g). We conclude with a brief summary.

2. Mass Evolution due to Hawking Radiation and Phantom Energy

Decrease in mass of a BH due to the combined effect of Hawking Radiation and phantom energy is given by [13]

$$\frac{dM}{dt} = \frac{dM}{dt}|_{hr} + \frac{dM}{dt}|_{p} \ , \tag{1}$$

$$\frac{dM}{dt} = -\frac{\hbar c^4}{G^2}\frac{\alpha}{M^2} + \frac{8G}{3c^3}\frac{M^2}{t_o^2}(1+\omega) \ , \tag{2}$$

where t_o is the time of domination of DE. Rewriting Eq. (2) as

$$\frac{dM}{dt} = -aM^2 - \frac{b}{M^2} \ , \tag{3}$$

where

$$a \equiv -\frac{8G}{3c^3 t_0^2}(1+\omega) \equiv \frac{8\epsilon G}{3c^3 t_0^2} \ , \tag{4}$$

$$b \equiv \frac{\hbar c^4 \alpha}{G^2} = 7.6804 \times 10^{25} g^3 s^{-1} \ . \tag{5}$$

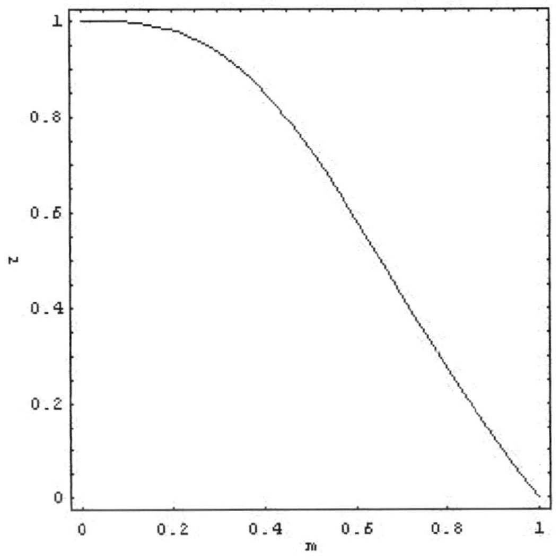

Figure 1. The normalized time is plotted against mass parameter for p=5.

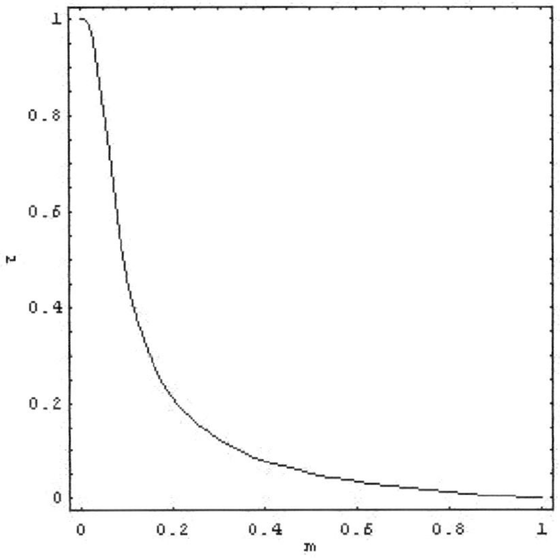

Figure 2. The normalized time is plotted against mass parameter for p=100000.

Eq. (3) can be written as

$$-\int dt = \frac{1}{b}\int \frac{M^2}{1+\frac{a}{b}M^4}dM \ , \tag{6}$$

$$\begin{aligned} t_f - t = & \frac{1}{4\sqrt{2}(a^3b)^{\frac{1}{4}}}[ln|\frac{1-\sqrt{2}(\frac{a}{b})^{\frac{1}{4}}M+(\frac{a}{b})^{\frac{1}{2}}M^2}{1+\sqrt{2}(\frac{a}{b})^{\frac{1}{4}}M+(\frac{a}{b})^{\frac{1}{2}}M^2}| \\ + & 2tan^{-1}(1+\sqrt{2}(\frac{a}{b})^{\frac{1}{4}}M) - 2tan^{-1}(1-\sqrt{2}(\frac{a}{b})^{\frac{1}{4}}M)] \ , \end{aligned} \tag{7}$$

where we have used the initial condition that black hole mass vanishes at t_f. By comparison of Eq. (7) with Eq. (24) of [13], the difference between the two can be seen. Here t =age of universe when mass of black hole is M, $M = mM_i$ (where $m = [0, 1]$). Thus Eq. (3) becomes

$$\frac{dm}{dt} = -a'm^2 - \frac{b'}{m^2} \ , \tag{8}$$

where $a' = aM$ and $b' = \frac{b}{M^3}$. Thus Eq. (7) becomes

$$\begin{aligned} t = t_f - & \frac{1}{4\sqrt{2}(a^3b)^{\frac{1}{4}}}[ln|\frac{1-\sqrt{2}(\frac{a'}{b'})^{\frac{1}{4}}m+(\frac{a'}{b'})\frac{1}{2}m^2}{1+\sqrt{2}(\frac{a'}{b'})^{\frac{1}{4}}m+(\frac{a'}{b'})^{\frac{1}{2}}m^2}| \\ + & 2tan^{-1}(1+\sqrt{2}(\frac{a'}{b'})^{\frac{1}{4}}m) - 2tan^{-1}(1-\sqrt{2}(\frac{a'}{b'})^{\frac{1}{4}}m)] \ . \end{aligned} \tag{9}$$

Defining

$$p = \frac{a'}{b'}, \tag{10}$$

where $a' = b'$ indicates equal contribution from phantom energy and HR in the evaporation of PBHs and normalizing Eq. (9), we get

$$t = t_f[1 - \frac{ln|\frac{1-\sqrt{2}p^{\frac{1}{4}}m+p^{\frac{1}{2}}m^2}{1+\sqrt{2}p^{\frac{1}{4}}m+p^{\frac{1}{2}}m^2}| + 2tan^{-1}(1+\sqrt{2}p^{\frac{1}{4}}m) - 2tan^{-1}(1-\sqrt{2}p^{\frac{1}{4}}m)}{ln|\frac{1-\sqrt{2}p^{\frac{1}{4}}+p^{\frac{1}{2}}}{1+\sqrt{2}p^{\frac{1}{4}}+p^{\frac{1}{2}}}| + 2tan^{-1}(1+\sqrt{2}p^{\frac{1}{4}}) - 2tan^{-1}(1-\sqrt{2}p^{\frac{1}{4}})}] . \tag{11}$$

To re-plot the graphs of [13], we take the normalized time $\tau = t/t_f$ and plot it against dimensionless mass parameter m. For $p \leq 1$ in Eq. (11), we get similar results, i.e., graphs as in [13]. But for $p > 1$, we get different graphs and hence their different interpretation. Figures $1 - 2$ are the plots for $p = 5$ and $p = 100000$ respectively. These graphs do not contain any non-physical or redundant part unlike figures 7 and 9 in [13].

3. Calculation of Transition Mass of PBHs

Eq. (10) can be written as

$$p = \frac{a}{b} M^4 . \tag{12}$$

$M\ |_{p=1} \equiv M_t$ is called the transition mass, i.e., the value of the mass of a PBH below which the evaporation due to HR dominates over the evaporation due to the phantom energy [14]. Using Eqs. (4) and (5), we get the transition mass

$$M_t = 3.4637 \times 10^{15} \frac{t_o^{\frac{1}{2}}}{\epsilon^{\frac{1}{4}}} g , \tag{13}$$

where t_o is the DE domination time measured in seconds. For $p < 1$, M is the corresponding upper bound for mass below which phantom energy is never dominant over HR; $p = 1$ gives the value of transition mass and $p > 1$ is the case when evaporation rate due to phantom energy is relatively more than that due to HR. Thus, we infer from here that figures $1 - 2$ have a part of the curve in the start which represents domination of phantom energy over the HR in the evaporation rate in a certain time interval.

It has been claimed on the basis of a double beta decay experiment that $0.05 < \epsilon < 0.67$ [17]. As such we evaluate M_t for the extreme and middle values and then include a much larger value ($\epsilon = 10$) to check the sensitivity of the transition mass to ϵ (in case the claim is invalid). For t_o to be 10 billion years, the results are given in table 1 for different values of ϵ.

Table 1. Mass of a PBH for Different Contribution from Phantom Energy for Different Values of ϵ, taking $t_o = 10$ billion years.

p	$M\ (g)\ (\epsilon = 0.05)$	$M\ (g)\ (\epsilon = 0.36)$	$M\ (g)\ (\epsilon = 0.67)$	$M\ (g)\ (\epsilon = 10)$
0.01	1.3011×10^{24}	0.7944×10^{24}	0.6801×10^{24}	0.3460×10^{24}
0.1	2.3138×10^{24}	1.4126×10^{24}	1.2094×10^{24}	0.6153×10^{24}
0.5	3.4599×10^{24}	2.1123×10^{24}	1.8085×10^{24}	0.9200×10^{24}
1.0	4.1146×10^{24}	2.5120×10^{24}	2.1508×10^{24}	1.0941×10^{24}
5	6.1527×10^{24}	3.7562×10^{24}	3.2161×10^{24}	1.6362×10^{24}
10	7.3168×10^{24}	4.4670×10^{24}	3.8246×10^{24}	1.9458×10^{24}

The values of M_t show that HR dominates over phantom energy for a sufficiently higher value of mass. Thus, the evaporation of a BH will not slow down towards the end as might have been expected from $dM/dt \propto M^2$, but will explode precisely as in the case of only HR. The only way to have the phantom energy effect significant is to reduce the value of M_t sufficiently, which is impossible since neither can we take much higher values for ϵ nor can we take much smaller values of t_o. Also, this implies that for a BH of initial mass less than M_t the dominant evaporation process is always HR.

For $t_o = 1$ billion years, we get the results shown in table 2. These results have a similar physical interpretation as the results of table 1. By comparing the results of both tables, we find that for a sufficiently early domination of DE, the HR will dominate over phantom energy at smaller values of mass (transition mass).

Table 2. Mass of a PBH for Different Contribution from Phantom Energy for Different Values of ϵ, taking $t_o = 1$ billion years.

p	$M\ (g)\ (\epsilon = 0.05)$	$M\ (g)\ (\epsilon = 0.36)$	$M\ (g)\ (\epsilon = 0.67)$	$M\ (g)\ (\epsilon = 10)$
0.01	0.4141×10^{24}	0.2512×10^{24}	0.2150×10^{24}	0.1094×10^{24}
0.1	0.7317×10^{24}	0.4467×10^{24}	0.3825×10^{24}	0.1946×10^{24}
0.5	1.0941×10^{24}	0.6680×10^{24}	0.5719×10^{24}	0.2910×10^{24}
1.0	1.3012×10^{24}	0.7944×10^{24}	0.6801×10^{24}	0.3460×10^{24}
5	1.9457×10^{24}	1.1878×10^{24}	1.0170×10^{24}	0.5174×10^{24}
10	2.3138×10^{24}	1.4126×10^{24}	1.2095×10^{24}	0.6153×10^{24}

4. Phantom Energy Accretion and Initial Mass of PBHs Evaporating Today

If we consider the evaporation due to HR only then by integrating the first term in Eq. (1) we get [1]

$$T = \frac{M_i^3}{3\,b}\,, \tag{14}$$

where M_i is the initial mass and T is the lifetime of a BH. This equation gives initial mass of a BH evaporating today (i.e. a lifetime of 4.3392×10^{17} s) due to HR to be $M_i = 4.64 \times 10^{14}$ g $\equiv M_{hr}$. Since the accretion of phantom energy accelerates the evaporation rate, a PBH of mass $\sim 10^{15}$ g will vanish earlier than today. If we take $t = time\ of\ formation\ of\ a\ PBH \equiv t_i$ then $M = M_i$. Thus Eq. (7) becomes

$$T \equiv t_f - t_i = \frac{1}{4\sqrt{2}(a^3 b)^{\frac{1}{4}}}[ln|\frac{1-\sqrt{2}(\frac{a}{b})^{\frac{1}{4}}M_i + (\frac{a}{b})^{\frac{1}{2}}M_i^2}{1+\sqrt{2}(\frac{a}{b})^{\frac{1}{4}}M_i + (\frac{a}{b})^{\frac{1}{2}}M_i^2}|$$
$$+\ 2tan^{-1}(1+\sqrt{2}(\frac{a}{b})^{\frac{1}{4}}M_i) - 2tan^{-1}(1-\sqrt{2}(\frac{a}{b})^{\frac{1}{4}}M_i)]\,. \tag{15}$$

Eq. (15) gives the lifetime of a PBH of some initial mass evaporating under the effect of HR and phantom energy. For physically reasonable values of $t_o \sim 1$ billion years, $a \sim 10^{-72}\ g^{-1}s^{-1}$ ($[a/b]^{\frac{1}{4}} \sim 10^{-26}$ g^{-1}). In fact, even for $t_o \sim 1$ s, $a \sim 10^{-39}\ g^{-1}s^{-1}$ ($[a/b]^{\frac{1}{4}} \sim 10^{-16}$ g^{-1}). As such, we can use a Taylor expansion for small $[a/b]^{\frac{1}{4}}M_i$. Thus we get

$$T = \frac{M_i^3}{3\,b}\,[1\ -\ \frac{1}{10}(\frac{a}{b})^{\frac{1}{2}}M_i^2\ +\ O(\frac{a}{b}M_i^4)]. \tag{16}$$

We then see that for $t_0 = 10$ billion years, $\epsilon = 0.1$ and $M_i = M_{hr}$, the correction factor is $\sim 10^{-22}$ as compared to 1 which corresponds to a decrease of $\sim 10^{-5}$ s in a lifetime of $\sim 10^{17}$ s! For $t_o \sim 1$ billion years, the correction factor is $\sim 10^{-21}$ which is also negligible as opposed to the claim in [13].

The lifetime of BHs of initial masses 10^{23} g, 10^{24} g and 10^{25} g are 4.34005×10^{42} s, 4.34005×10^{45} s and 4.34005×10^{48} s respectively if evaporation only due to HR is considered. When effect of phantom energy accretion is also considered in evaporation process (taking $t_o = 10$ billion years), Eq. (15) gives the lifetime to be 4.34005×10^{42} s, 4.33998×10^{45} s and 3.28317×10^{48} s in the same order. These results show that the effect of phantom energy accretion in evaporation process is significant for such higher masses.

Also, for $\epsilon = 0.1$ the transition mass is $\sim 10^{24}$ g which is greater than M_{hr} by 10 orders of magnitude. We conclude that for a PBH evaporating today under the combined effect of HR and phantom energy, the dominant evaporation process is HR even for a very early domination of phantom energy.

5. Summary

There were algebraic errors in [13] that have been corrected in this paper. The correction makes a significant difference. First, there was nonphysical behavior appearing in the graphs for the variation of mass with time that caused severe problems of interpretation. With the correction that nonphysical behavior has disappeared.

More importantly, it is found that due to the accretion of phantom energy the lifetime of a BH is decreased, but not by much. More substantial analytical work has been done to find the first order correction term in the lifetime of a PBH of some initial mass evaporating under the combined effect of phantom energy accretion and HR. There had been a claim of an enormous effect of phantom energy [13] whereby the required initial mass changed by 10 orders of magnitude if the initial phantom energy is 10 times the HR. We found that the increase in initial mass of a PBH evaporating today under the combined effect of phantom energy accretion and HR is insignificant.

Further, we have discussed the transition mass and studied its variation with EoS parameter ($\omega < -1$) numerically. We found that M_t came out to be larger than M_{hr} by approximately 10 orders of magnitude. This shows that phantom energy cannot be more significant in the evaporation process of a PBH evaporating today than Hawking radiation. In fact, for it to become equally significant we would have to wait 10^{31} times the present age of the Universe!

Acknowledgments

We would like to thank Maqbool Ahmed and Mubasher Jamil for useful comments and suggestions. One of us (SN) is grateful to NUST and AS-ICTP for providing full support to participate in the Third Italian-Pakistani Workshop on Relativistic Astrophysics, Lecce, Italy.

References

[1] Hawking S W 1974 *Nature* **248** 30
[2] Smirnov A I 2007 *Russian Phys. Journal* **50**
[3] Carr B J 2005 *Proc. of " Inflating Horizon of Particle Astrophysics and Cosmology"*
[4] Hubble E 1929 *Proc. Nat. Acad. Sci.* **15** 168
[5] Wiltshire L D 2008 *Int. Journal of Modern Phys.* D **17** 641-49
[6] Mukhanov V 2005 *Physical Foundations of Cosmology* (Cambridge University Press) p 263
[7] Coles P and Lucchin F 2002 *Cosmology, The Origin and Evolution of Cosmic Structure* (John Wiley and Sons, Ltd.) pp 119-29
[8] Copeland E J, Sami M and Hinji S 2006 *Int. J. Mos. Phys.* D **15** 1753-1847
[9] Caldwell R R 2002 *Phys. Lett.* B **545** 23
[10] Caldwell R R, Kamionkowski M M and Weinberg N N 2003 *Phys. Rev. Lett.* 071301

[11] De Paolis F, Jamil M and Qadir A 2010 *Int. J. Theor. Phys.* **49** 621-32
[12] Babichev E, Dokuchave V and Eroshenko Yu 2004 *Phys. Rev. Lett.* **93**
[13] Jamil M and Qadir A 2010 *Gen. Relativ. Gravit.* **43** 1069-82
[14] Guariento D C, Horvath J E, De Freitas Pacheco J A and Custodio P S 2008 *Gen. Relativ. Gravit.* **40** 1593-1602
[15] Nayak B and Singh L P 2011 arXiv:0905.3243v2 [gr-qc]
[16] Nayak B and Jamil M 2011 arXiv:1107.2025v1 [gr-qc]
[17] Macorra A D L, Melchiorri A, Serra P and Bean R 2007 *Astropart. Phys.* **27** 406-410
[18] Carr B J 1975 *Ap. J.* **201** 1
[19] Hawking S W 1971 *Phys. Rev. Lett.* **26** 1344
[20] Carr B J and Hawking S W 1974 *MOn. Not. R. Astr. Soc.* **168** 399
[21] Page D N 1976 *Phys. Rev.* D **13** 198
[22] Adams F C, Mbonye M and Laughlin G 1999 *Phys. Letts.* B **450** 339-42
[23] Grillo A F and Srivastava Y 1981 *Nuclear Physics* B **187** 271-84
[24] Grillo A F 1980 *Phys. Letts.* B **94** 364-66
[25] Petkov V B et al 2008 *Astron. Tett.* **34** 509-14
[26] Petkov V B, Bugaev E V, Klimai P A and Smirnov D V 2008 *JETP Lett* **87** 1-3

Journal of Physics: Conference Series **354** (2012) 012013 doi:10.1088/1742-6596/354/1/012013

X-rays from eclipsing cataclysmic variable systems: the eclipse profile

A.A. Nucita

Department of Physics, University of Salento, Via per Arnesano, CP 193, I-73100, Lecce, Italy& INFN, Sez. di Lecce, via Per Arnesano, CP 193, I-73100, Lecce, Italy

E-mail: nucita@le.infn.it

F. De Paolis

Department of Physics, University of Salento, Via per Arnesano, CP 193, I-73100, Lecce, Italy& INFN, Sez. di Lecce, via Per Arnesano, CP 193, I-73100, Lecce, Italy

E-mail: depaolis@le.infn.it

G. Ingrosso

Department of Physics, University of Salento, Via per Arnesano, CP 193, I-73100, Lecce, Italy& INFN, Sez. di Lecce, via Per Arnesano, CP 193, I-73100, Lecce, Italy

E-mail: ingrosso@le.infn.it

Abstract. A cataclysmic variable is a binary system containing a white dwarf which accretes material from a secondary star through the Roche lobe mechanism. Systems observed at large inclination angles offer the possibility to locate the the X-ray emitting region thus allowing to study the behaviour of systems accreting matter in extreme conditions. We briefly review the main properties of a cataclysmic variable and focus on the information possibly derived by high energy observations.

1. Introduction

A binary system constituted by a white dwarf (the primary star) gravitationally interacting with a secondary object (the secondary star) is called Cataclysmic Variable (hereafter CV).

In CV systems, the two stars interact via the formation of a Roche-lobe, and the details of the accretion are governed by many parameters. In particular, depending on the magnetic field, these binary systems are classified as: (i) non-magnetic systems, with a rather weak magnetic field of strength $\lesssim$0.1 MG), characterized by a keplerian accretion disk around the compact object; (ii) intermediate polars with magnetic field of the order 0.1-10 MG, where the disk is trucanted close to the white dwarf; (iii) polars, i.e. highly magnetized systems with $\gtrsim$10 MG, where the accretion occurs via a ballistic jet which hits one of the poles of the primary star.

In a typical non-magnetic CV, half of the potential gravitational energy of the accreting material is dissipated by viscosity, while the remaining part is radiated away by the boundary layer (BL), i.e., a region at the interface between the disk and the white dwarf surface. In particular, as demonstrated by [1], the total luminosity of the accretion disk of a white dwarf of mass M_{WD} and radius R_{WD} is $L_{disk} = GM_{WD}\dot{M}/2R_{WD}$. Thus, only half of the gravitational potential energy $L_{acc} = GM_{WD}\dot{M}/R_{WD}$ is radiated away by the accretion disk, while the other half is released in the form of rotational energy in the accreting material. A fraction of this energy is dissipated at the BL, where the velocity of the accreting gas decreases, with a rate depending on the rotational speed of the white dwarf. Taking into account the energy transferred to spin up the white dwarf, the boundary layer luminosity is $L_{BL} \simeq 0.125 L_{acc}$ for $\Omega_{WD} \simeq 0.5\Omega_K(R_{WD})$ [1]. Thus, the radiation emitted by the disk reaches a maximum in the optical and ultraviolet spectral regions, while the BL, if exists, may radiate in the extreme ultraviolet and X-rays, with typical luminosities in the range 10^{30}–10^{32} erg s^{-1} (see e.g., [2], [3]).

The low luminosity of these systems implyies that the *XMM*-Newton satellite [4] is particularly efficient in their study since it has a large effective area and offer the possibility to observe the source simultaneously in the optical band with the onboard optical monitor (OM).

Even if according to the standard accretion scenario the X-ray photons are expected to originate in the BL, what present observations clearly show is that the high energy emission comes from a region very close to the white dwarf. Moreover, no clear correlation between the X-ray luminosity and the accretion rate does exist [5]. In addition, since in many CVs the observed X-ray luminosity is lower than predicted by the BL model, the details of the accretion mechanism are still unclear (but see e.g., [6] for possible solutions about the observed discrepancy).

Systems characterized by large inclination angles with respect to the line of sight offer the unique possibility to study the location of the X-ray emitting region. In fact, under these circumstances, the observed X-ray light curve shows the existence of peculiar features as eclipses whose detailed study allows to constrain the X-ray source size when the mid ingress and egress phases of the occultation are known. Of course, the main difficulty in this kind of researches is that the typical low X-ray luminosity of a CV becames even lower during an undergoing eclipse phenomenon.

A first notable example of this kind is the well studied dwarf nova system OY Carinae which is nowadays considered as the prototype dwarf novae. In OY Car, the X-ray eclipse was resolved by the *XMM*-Newton satellite, and, as shown in [7], the size of the X-ray emitting region is comparable with that of the central white dwarf. In addition, a careful re-analysis of the same data by [8] allowed to depict a scenario in which the X-ray emission is probably displaced from the equatorial region. Hence the authors concluded that the X-ray emission in the dwarf nova OY Car could possibly originate at the white dwarf polar cap (see their Fig. 5), so that magnetic accretion may play an important role also in those systems formally recognized as non magnetic ones. However, as the same authors also suggested, the obscuration of the lower WD hemisphere by the inner accretion disk would be a valid alternative model implying that the observed X-ray properties are caused only by projection effects.

In the following we will review the main porperties of the high energy light curves of two interesting dwarf novae systems: HT Cassiopea and Z Chamaleontis and show how the size of the X-ray source can be determined from observation.

2. The folded light curves of HT Cassiopea and Z Chamaleontis: two interesting eclisping dwarf novae systems

HT Cassiopea and Z Chamaleontis (hereafter HT Cas and Z Cha, respectively) are interesting targets for the *XMM*-Newton satellite for several reasons[1]: **a)** they are characterized by high inclination angles (81^0-82^0) with respect to the line of sight, **b)** have orbital periods of the order

[1] For further details on the data reduction scheme, the spectral as well as the timing analysis of the two sources we address the reader to [9] and [10], respectively.

of a few hours ($\simeq 1.77$ h -[11]- and $\simeq 1.78$ h -[12]-), and **c)** emit a sufficient large amount of X-ray photons ($L_X^{Bol}s \simeq 7 \times 10^{30}$ erg s^{-1} - 1×10^{31} erg s^{-1}). It is then possible to observe for a sufficiently long time the two luminous CVs and infer the X-ray source size after characterizing the eclipse profile.

HT Cas was observed by *XMM*-Newton on two occasions in 2002 (Observation IDs 0111310101 and 0152490201), for $\simeq 45$ ks and $\simeq 55$ ks, respectively. Here, we refer only to the first observation corresponding to the normal quiescent state of the system. The observation was conducted on 2002/08/20 and started (ended) at 09 : 23 : 50.0 (23 : 14 : 42.0) UT.

Similarly, Z Cha was observed by *XMM*-Newton in 2003 and 2005 (Observation IDs 0205770101 and 0306560301) for $\simeq 99$ ks and $\simeq 87$ ks, respectively. Also in this case, we focused on the X-ray eclipse of Z Chamaleontis so that we used only the observation (ID 0205770101) clearly showing eclipses in the light curves. The observation started on 2003 December 19 (20 : 45 : 52.0 UT) and ended on 2003 December 21 (00 : 56 : 0.9 UT).

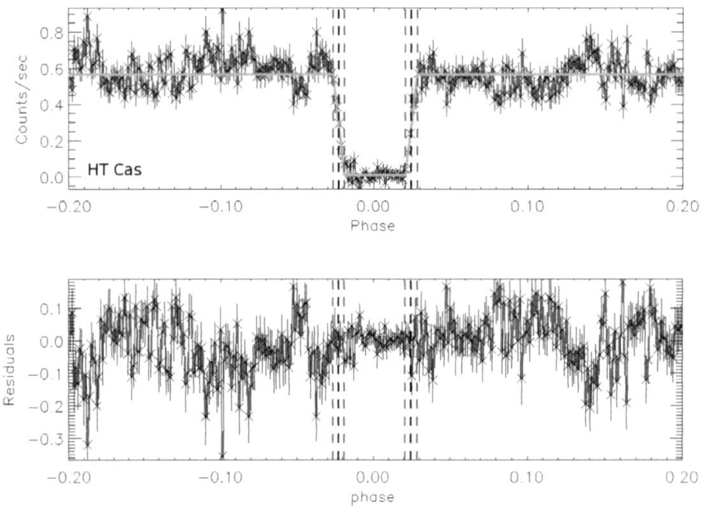

Figure 1. A piecewise fit (solid green line) to the HT Cas EPIC folded light curve (upper panel), and the residuals (as the difference between the data and the best fit model).

In Figures 1 and 2, we give the high energy light curves (in the energy range 0.5-10 keV band) for HT Cas and Z Cha, respectively. The two light curves were folded at the associated orbital period and two periods are shown clearly demonstrating the existence of deep eclipses. Here (but see also [9] and [10] for details), the light curves are fitted via a phenomenological model consisting of a constant level (outside the eclipse), a flat part inside the eclipse, and two linear ingress and egress functions (not necessarily equal). Thus, after adapting to the data the 6 free parameters via a minimization of the χ^2 function it is possible to determine the so called contact points, i.e. the phases corresponding to which the X-ray emitting region disappears and reappears from behind the eclipsing source (the secondary star). In more details, the X-ray emitting region disappears between phases ϕ_1 and ϕ_2, and reappears between phases ϕ_3 and ϕ_4, so that the durations of the ingress and egress are $\Delta\phi_{1,2} = |\phi_2 - \phi_1|$ and $\Delta\phi_{3,4} = |\phi_4 - \phi_3|$, respectively. Furthermore, the mid ingress ϕ_i and mid egress ϕ_e points (i.e. the phases corresponding to which half of the light curve value is eclipsed) are used to estimate the eclipse duration $\Delta\phi_X = |\phi_e - \phi_i|$. These phases are represented in the figures by dashed vertical lines.

The contact point determined by using the X-ray folded light curve can be used in order

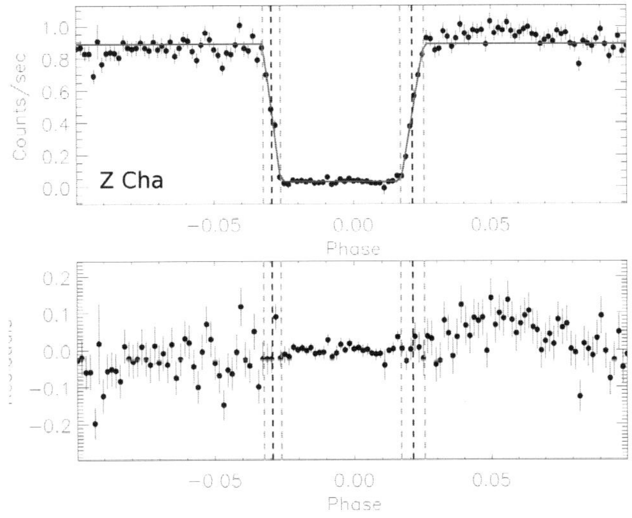

Figure 2. The same as in figure 1 but for the Z Cha binary system.

to estimate the size the X-ray emitting region. Assuming that the cross sections of both the primary star (with radius r_2) and the secondary star (with radius r_1) remain constant over the eclipse duration, the timescales for the eclipse ingress/egress can be evaluated as [14]

$$\Delta t_{1,2} = \frac{P}{2\pi}\left[\sqrt{\left(\frac{r_1+r_2}{a}\right)^2 - \cos^2 i} - \sqrt{\left(\frac{r_1-r_2}{a}\right)^2 - \cos^2 i}\right]. \quad (1)$$

where P, a, and i are the orbital period, the semi-major axis and inclination angle, respectively. Note that the phenomenological model used above can be substituted by a more realistic

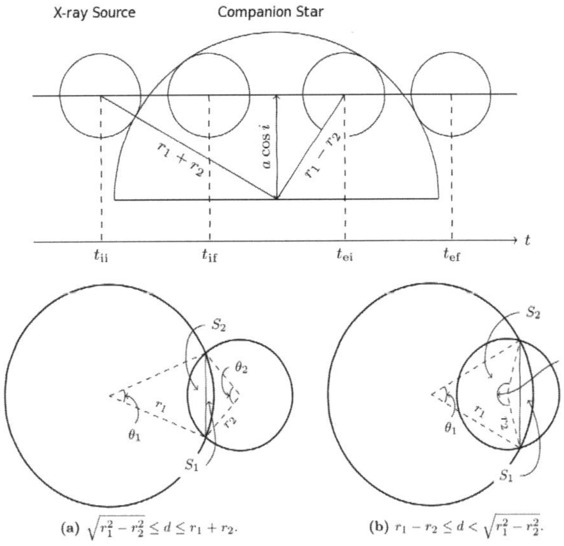

Figure 3. In the figure, we show the geometry of an undergoing eclipse event and quote all the relevant variables used in the text.

alharacterization of the eclipse profile. Here, one assumes that the limb of the companion star is completely opaque to the X-ray photons during an eclipse event. In particular, referring to Figure 3 (bottom panels), one can easily understand that the eclipse profile depends on the fraction of the area ($\delta A = S_1 + S_2$) of the X-ray source covered (at a given time t) by the limb of the companion star. It is straightforward to verify that the area δA is

$$\delta A(t) = \begin{cases} 0, & \text{for } d(t) > r_1 + r_2, \quad (2) \\ \frac{r_1^2}{2}(\theta_1(t) - \sin\theta_1(t)) + \frac{r_2^2}{2}(\theta_2(t) - \sin\theta_2(t)), & \text{for } r_1 - r_2 \leq d(t) \leq r_1 + r_2, \quad (3) \\ \pi r_2^2, & \text{for } d(t) < r_1 - r_2. \quad (4) \end{cases}$$

where the projected distance $d(t)$ between the centers of the X-ray source and the companion star can be evaluated as

$$d^2(t) \simeq 4\pi^2 a^2 \phi^2(t) + a^2 \cos^2 i, \quad (5)$$

where $\phi(t)$ is the orbital phase and

$$\theta_1(t) = 2 \arccos \frac{r_1^2 - r_2^2 + d^2(t)}{2r_1 d(t)}, \quad (6)$$

$$\theta_2(t) = 2 \arccos \frac{r_2^2 - r_1^2 + d^2(t)}{2r_2 d(t)}. \quad (7)$$

Thus, the resulting eclipse profile (e.g. count rate) is

$$L(t) = L_0 \left(1 - \frac{\delta A}{\pi r_2^2}\right), \quad (8)$$

where L_0 is the X-ray count rate outside the eclipse phenomenon.

Using the piecewise phenomenological model and/or the more realistic model described above together with the measured X-ray ingress/egress timescales and the orbital parameters determined by the optical observation (i.e., P, a, i, and r_1), one can derive the size the HT Cas and Z Cha X-ray emitting regions to be $\simeq 0.0117$ R$_\odot$ and $\simeq 0.0119$ $R_\odot$, respectively. Since these values (and the associated errors) are comparable with the WD radii estimated by using the optical observations (see [13] and [15]), the BL model seems to be enforced.

References

[1] Popham, R., & Narayan, R., 1985, ApJ, 442, 337.
[2] Lamb, D.Q., 1982, *Cataclysmic Variables and Related Objects: proceeding of the 72nd colloquium of the international astronomical union*, (Eds. Livio, M. & Shaviv, G.).
[3] Baskill, D.S.., Wheatley, P.J., & Osborne, J.P., 2005, *MNRAS*, **357**, 626.
[4] Jansen, F., et al., 2001, *A&A*, **365**, L1.
[5] van Teeseling, A., Beuermann, K., & Verbunt, F., 1996, *A&A*, **315**, 467.
[6] Ferland, G. J., et al., 1982,*ApJ*, **262**, L53.
[7] Ramsay, G., 2001, *A&A*, **365**, L288.
[8] Wheatley, P.J., &, West, R.G., 2003, *MNRAS*, **345**, 1009.
[9] Nucita, A.A., Maiolo, B.M.T., Carpano, S., Belanger., G., et al., 2009, *A&A*, **504**, 973.
[10] Nucita, A.A., Kuulkers, E., Maiolo, B.M.T., De Paolis, F., et al., 2011, *A&A*, in press.
[11] Borges, B.W., Baptista, R., Papadimitriu, C., & Giannakis, O., 2008, *A&A*, **480**, 481.
[12] Baptista, R., Jablonski, F., Oliveira, E., Vrielmann, S., et al., 2002, *MNRAS*, **335**, L75.
[13] Horne, K., Wood, J.H., & Stiening, R.F., 1991, *ApJ*, **378**, 271.
[14] Wheatley, P.J., &, West, R.G., 2002, *The Physics of Cataclysmic Variables and Related Objects, ASP Conference Series*, Vol. 261, (Eds. Gänsicke, B.T., Beuermann, K. & Reinsch, K.).
[15] Wood, J. H., Horne, K., Berriman, G., Wade, R., et al. 1986, *MNRAS*, **219**, 629.

Journal of Physics: Conference Series **354** (2012) 012014 doi:10.1088/1742-6596/354/1/012014

Self-interaction of gravitational waves and their observability

Asghar Qadir

Center for Advanced Mathematics and Physics, National University of Sciences and Technology, H-12, Islamabad, Pakistan

E-mail: aqadirmath@yahoo.com

Abstract. Energy is not a well-defined concept in General Relativity (GR). There have been various approaches adopted, including simply discarding the concept. However, with mass-energy equivalence and the importance of mass in GR, many have thought it worth preserving. Of the various suggestions for defining energy, there is a proposal that the energy content in a time varying spacetime can be obtained by considering its second approximate Lie symmetries. Unlike others, this proposal allows one to calculate the energy in gravitational waves. When applied to cylindrical gravitational waves, it gives a self-damping of the waves. Though other waves may be expected to have different mathematical behaviour, the physics of self-damping should remain the same and one would expect a qualitatively similar result, even if it is quantitatively different in some details. If this proposal is valid, the estimates for the required sensitivity of the detectors for the various sources will have to be revised upwards. This revision is worked out here. In view of the fact that after two years of being online no signal has been seen by the detectors, it is argued that this proposal should be taken as a serious contender for a valid definition of energy in GR.

1.. Introduction

In the early days of Relativity the reality of gravitational waves was questioned as they are solutions of the vacuum field equations and hence apparently have no energy. Weber & Wheeler [32] demonstrated that they would impart momentum to test particles in their path. This demonstration was extended to test particles in the path of plane gravitational waves [5]. Later, a general closed formula for the momentum imparted to test particles in an arbitrary spacetime was given [24].

Though it seemed obvious that "the waves must carry energy which is to be imparted", no clear measure for the energy they carried was available. This is because energy is not generally conserved in GR and hence mass and energy are not defined. One way to avoid this problem is by defining a stress-energy "pseudo-tensor" (see for example [17]), but it is observer dependent and hence not generally accepted by relativists for defining the energy (see for example the discussion of the pseudo-tensor in [21]). A deeper approach is to define some measure for the breaking of time translational symmetry as a measure for the energy density in the spacetime [3, 12, 15, 20, 25, 27, 33]. In that sense, the energy is "hidden" in the field. None of these attempts showed unambiguous success as they did not lead to any basic new physical insights or predictions. One would have hoped that a "correct" interpretation would announce itself by leading to a testable prediction. This hope implicitly regards GR as incomplete. Many would argue that it is so well tested that it should be accepted as complete, except in the quantum domain. I take the view that a problem cannot be discarded when it is inconvenient and should be faced "square-on" as Wheeler used to say. It could even be that the problems of the quantum may be elucidated by a completion of the theory at a classical level.

Following Wheeler's approach of confronting the problem with any physical theory at its severest, nowhere is the problem of energy in GR as severe as in gravitational waves, where we have lost time translational symmetry and yet have no stress-energy tensor. Unlike electromagnetic waves (which do not undergo self-interaction because the Maxwell theory is linear), gravitational waves undergo self-interaction due to the non-linearity of GR. This self-interaction could, in principle, cause damping, like Landau damping of electromagnetic waves due to their interaction with matter [16], or cause an *enhancement* such as might be expected on the basis of the work on colliding gravitational waves [14, 26], where the spacetime curvature increases in some regions so that singularities are formed due to the interaction between two plane gravitational waves.

One approximate symmetry proposal is the use of broken (or approximate) Lie symmetries [1, 11] of the geodesic equations [10]. The breaking is taken to be by terms involving a small parameter whose powers, higher than some chosen value, can be neglected. It was found that though with the first order approximate symmetries no new insight is obtained [13], for the second order the timelike symmetry picks up a scaling factor [7, 8]. As this factor does not depend on the strength of the breaking it can be taken to be zero and hence the form of the perturbation should be irrelevant (much as displacements are inserted and then taken to be zero for calculations in Statics using D'Alembert's principle). Thus, it may provide a genuine measure for the energy content of gravitational waves.

This proposal has been implemented for cylindrical gravitational waves [9]. For them the scaling factor turns out to go asymptotically as

$$f(\rho,\omega) \sim \frac{3\times 2^{11/4}}{\pi^{3/2}}[(|\cos(\omega\rho/c)|)^{3/2}\sin(2\omega t)](\omega\rho/c)^{-1/2} + O((\omega\rho/c)^{-3/2}), \tag{1}$$

where ρ is the radial distance from the source and ω is the frequency of the wave. This formula does not give the energy content of the waves but only the *scaling* of the energy expected. The energy content estimate must come from the classical expectation, which is also recovered by first order approximate Lie symmetry analysis [10, 13]. This scaling gives the effect of the self-interaction of the waves due to their nonlinearity as a *damping* and not an enhancement.

The self-damping of gravitational waves will obviously have direct observational consequences. Waves that should just have been detectable on the original expectations would go below the level of sensitivity of the detector. In view of the fact that the waves have not been detected for two years that the detectors have been online, as of the middle of 2011 [18], it seems worth taking the proposed definition of energy seriously. As will be seen, the effect of self-damping is very significant. There have been objections raised that it is unbelievable that such strong effects could occur in what is a linearizable theory. My point is that it is only an assumption that the theory *is* linearizable in the regime under consideration. That it be linearizable requires that the nonlinear effect be negligible compared with the linear. Since there is nothing to compare with, there is no basis for that assumption.

The plan of the paper is as follows. In the next section I will give a very brief introduction to approximate Lie symmetry analysis and review the consequences deduced from it for our purposes. In section 3 I will explore the observational significance of the proposal. In the subsequent section I will give a "poor-man's way" (a'la Wheeler) of trying to interpret the physics of the effect. In the concluding section I will give a brief summary of the arguments and discuss them. It will be argued that the broad effect is inevitable but the specific values for the self-damping depend on the proposal and can only rely on observations for validation.

2.. Approximate Lie symmetry analysis

Given a system of p partial differential equations (PDEs) of order m for n functions, s^a of q variables, t^i

$$E_\alpha(t^i, s^a; s^a_i, s^a_{ij}, ..., s^a_{i...m}) = 0 \ , \tag{2}$$

we treat it as an *algebraic* equation for the independent and dependent variables and their partial derivatives up to order m. The equation is said to be symmetric under a point transformation $(t^i, s^a) \rightarrow (x^i, y^a)$ if it results in the same algebraic expression in terms of the new variables. If there is only one independent variable, t, it is a system of ordinary differential equations (ODEs). We shall, here, only be interested in ODEs. For these equations we shall denote the derivatives by s'^a. Lie [19] showed that if there are enough symmetries then the system of equations can be solved by transformation of variables.

Since one is dealing with differential equations one needs that the functions be differentiable of the relevant order. As such, one uses differentiable transformations and hence we can consider infinitesimal generators of symmetry. The generic generator is

$$\mathbf{X} = \xi(t, s^a)\frac{\partial}{\partial t} + \eta^b(t, s^a)\frac{\partial}{\partial s^b} \ , \tag{3}$$

where ξ and η^a are arbitrary functions. Of course, we need to extend the generator to incorporate the derivatives of the dependent variables. For a second order system of ODEs,

$$\mathbf{X}^{[1]} = \xi(t, s^a)\frac{\partial}{\partial t} + \eta^b(t, s^a)\frac{\partial}{\partial s^b} + \eta^{1b}(t, s^a)\frac{\partial}{\partial s'^b} + \eta^{2b}(t, s^a)\frac{\partial}{\partial s''^b} \ , \tag{4}$$

where

$$\eta^{a[n]} = \frac{d\eta^{a[n-1]}}{dt} - s^{an}(r)\frac{d\xi}{dt} \ , \tag{5}$$

$(\eta^{a[0]} = \eta^a)$, d/dt stands for the total derivative,

$$\frac{d}{dt} = \frac{\partial}{\partial t} + s'^b\frac{\partial}{\partial s^b} + s''^b\frac{\partial}{\partial s'^b} \tag{6}$$

and $s^{na}(t)$ stands for the nth derivative of s^a relative to t. Then

$$\mathbf{X}E_a|_{E_a=0} = 0 \ , \tag{7}$$

i.e. the symmetry generators, acting on the system of equations will be zero when evaluated for the solutions of the given system of equations.

If the above equation is *not* satisfied but if

$$E_a = E_a^0 + \epsilon E_a^1 + \epsilon^2 E_a^2 + ... \tag{8}$$

and there exist an infinitesimal symmetry generator, $\mathbf{X}_0$, of E_a^0, and $\mathbf{X}_1$ such that

$$\mathbf{X} = \mathbf{X}_0 + \epsilon \mathbf{X}_1 \tag{9}$$

and

$$\mathbf{X}E_a|_{E_a=0} = O(\epsilon^2) \ , \tag{10}$$

then we say that $\mathbf{X}$ is an *approximate* symmetry generator of the system of equations $E_a = 0$. If $\mathbf{X}_0 = k\mathbf{X}_1$, with k allowed to be zero or a non-zero constant then $\mathbf{X}$ is said to be a *trivial* approximate symmetry. If that is not the case then it is said to be *non-trivial.*

Higher order approximate symmetries can be obtained similarly. In particular we can define *second approximate symmetries* by asking for

$$\mathbf{X} = \mathbf{X}_0 + \epsilon \mathbf{X}_1 \epsilon^2 \mathbf{X}_2 \ , \tag{11}$$

now with the requirement that

$$\mathbf{X}E_a|_{E_a=0} = O(\epsilon^3) \ . \tag{12}$$

While for (first) approximate symmetries it makes no difference whether we use solutions of $E_a = 0$ or $E_a^0 = 0$, for higher order symmetries in general *it does.* This fact will be important.

The infinitesimal symmetry generators for the system of ODEs form a Lie algebra. The symmetries for a system of ODEs for a free particle in n-dimensional space admit the Lie algebra $sl(n+2, \mathbb{R})$. For Minkowski spacetime there are, then, 10 isometries and 35 generators. As soon as the isometries of the space are reduced the Lie symmetries are also (drastically) reduced. For Schwarzschild spacetime there are only 4 isometries (corresponding to angular momentum and energy conservation) and only 6 Lie symmetries (the 4 isometries and the generators for geodetic re-parametrization). The extra symmetries are $\partial/\partial s$ and $s\partial/\partial s$, where s is the geodetic parameter, i.e. *the proper time.* The extra isometries, corresponding to momentum and spin conservation, are recovered as (first) approximate symmetries [13]. (The radial equation of motion also has two non-trivial approximate symmetries, but that is not relevant for our present purpose.) However, the Reissner-Nordström spacetime does not possess *any* (first) approximate symmetries (or any non-trivial approximate symmetries for the radial equation). To further extend the analysis used for the Schwarzschild spacetime, we tried using second approximate symmetries [7, 8]. We found that we could recover the lost conservation laws as (second) approximate symmetries *provided that we re-scaled the proper time translation symmetry generator by a position dependent factor.* Since time translation corresponds to energy, the factor gives a position-dependent re-scaling of energy. The results for the Reissner-Nordström [7] and the Kerr spacetimes [8] gave physically reasonable results.

To follow through the programme we then applied the procedure with second approximate symmetries to the exact cylindrical gravitational wave spacetime. The result was a position dependent re-scaling of the energy of the cylindrical wave [9]. This was interpreted as giving the change of energy due to self-interaction. The factor involves a Bessel function of order zero. The asymptotic behaviour of this scaling factor, given in Eq.(1), is to *reduce* the energy by a factor of a square-root of the distance from the axis.

3.. Observational implications

The main purpose of this paper is to examine the observational implications of this attenuation of gravitational waves implicit in the approximate symmetry proposal and look for predictions that can be tested.

The obvious implication of the predicted effect is that our expectation of how the energy in the wave decreases with the distance of the source and the frequency of the wave (and hence the length of the antenna) will be very seriously overestimated. However, one can not simply take the attenuation factor to be $\sqrt{c/\rho\omega}$, as that is found for cylindrical waves. There are no infinite axially symmetric sources in Nature to emit genuine cylindrical waves. Nor is it a priori clear that a spherical, or slightly aspherical, source will give the same attenuation factor.

Since a cylinder is "*less* curved" than a sphere (in that the cylinder can be constructed from a plane sheet, while the sphere cannot) the curvature effect of a spherical wave should not be less than that of a cylindrical wave. The classical spherical wave energy density $E_{cs} \sim 1/\rho^2$. We would thus expect that $E_{sph} \sim \rho^{-5/2-\alpha}, \alpha > 0$. The cylindrical waves depend on the distance, via a Bessel function of order zero. The asymptotic approximation of this function gives the scaling factor mentioned above. One would expect that the energy in spherical waves will depend on the distance via a spherical Bessel function, which has the same asymptotic behaviour. As such, the scaling factor can be expected to be the same for both types of waves. Of course, there are no exact spherical gravitational waves either. However, we could approximate the effect of the spherical waves by patching together two Nutku "spherical waves with strings attached" [22]. Ideally, one would like to know exactly how much the attenuation for realistic models of the source. This analysis is vital for a more precise test of the proposal, but we will use the weaker condition for our present purposes. Here we will just take the least attenuation that there can be, namely that of the cylindrical waves.

The attenuation factor needs to be considered more carefully from the observational viewpoint. The frequencies observed will depend on the length of the detector, in that the waves must "fit into them". Thus the wavelengths must be simple fractions of the detector length. The series for $1/\sqrt{n}$ has very slow convergence. However, the spectrum of the waves may give faster convergence. Unfortunately, we cannot use the Planck spectrum for the harmonics as it derives from the quantum nature of the electromagnetic field, which is not directly applicable to gravitational waves. Since quantum gravity only applies at ultra-high energies, the relevant spectrum can be estimated by an exponentially damped tail. Roughly estimating the attenuation factor for this exponential damping, gives 1.5 times the value for the basic mode, which has a wavelength half of the detector length.

An interesting source to consider for our purpose is SN1987A, which had a 0.1 asphericity in its explosion [28]. One could regard the resulting waves as, in some sense, combined "spherical" and cylindrical waves. Crudely, then, one could take a cylindrical component of 0.1 of the total and assume that it has the predicted attenuation factor. The axis of the explosion is estimated to make an angle of $\pi/4 = 45^o$ with our line of sight. It seems reasonable to estimate the warped geometry attenuation factor $f \sim \sqrt{c/\rho^{3+\alpha}\omega}$. This reduction is enormous for SN1987A, which is at a distance of ~ 51.4 *kpc*. For an interferometer with an arm of ~ 3 *km* the attenuation factor comes out to be about a million! For a bar detector it would be *greater* by a factor of about twenty. For the [6] binary pulsar F would be about 300,000 for the interferometer and a few millions for the bar detector.

4.. Physical interpretation of the scaling

Notice that $\omega\rho/c$ is the number of waves, n, that fit into the distance ρ. Thus $f \sim 1/\sqrt{n}$. It is worth trying to see this in Wheeler's "poor man's way" to fully understand what is happening. Whereas the Newtonian space and time can be thought of as a rigid spatial grid and an orthogonal rigid line of time, and the special relativistic spacetime as a rigid grid of space and time the general relativistic spacetime is a flexible grid of space and time. The former can be thought of as made of steel girders and the latter of rubber girders. The wave would then travel through the warped spacetime. As such, the distance traveled would not be the classically estimated distance but distance *along* the "troughs" and "peaks" of the wave. Another way of visualizing it is an oscillating membrane (like a trampoline). The length along the membrane at maximum stretching will be more than the diameter. Hence we should anyhow expect the effective distance that the wave travels to be greater than the classical distance, and hence the wave to be attenuated relative to the classical expectation. Further, the attenuation should depend on n. The question remains, exactly how it depends on n. The proposed definition of energy gives the factor noted above.

For classical cylindrical waves the energy density $E_{cc} \sim 1/\rho$. (In terms of the proposal, this energy is recovered by the first approximate symmetries.) Thus the total energy density of the waves is asymptotically $E_{cyl} \sim \sqrt{c/\rho^3\omega}$. This result runs counter to our usual intuition based on a linear field like electromagnetism. In that case the loss of energy would have to be compensated by a gain of energy in the intervening region, causing heating of that region. Here that energy has gone into warping, and thus enlarging the spacetime. Again, the dependence on the frequency of the wave seems counter-intuitive from the linear view-point but is easily seen to be reasonable from the warped spacetime viewpoint.

5.. Conclusion

Weber claimed to have observed a signal on his *non*-cryogenic bar detector synchronous with the light and neutrino burst from SN1987A [29]. Since the standard theory of the bar detector developed by him [30] leads to energy requirements incompatible with the energy output of SN1987A ($\sim 10^{53}$ *ergs*), this claim was generally regarded as untenable. Indeed, Weber tried to revamp his theory of the detector to provide much greater efficiency of detection, using a claimed quantum coherence effect [23, 31]. While Bassan did not believe the claim, he admitted to having seen the signal claimed by Weber, as he was visiting Weber's laboratory at the time [2]. If our proposal is borne out, then Weber's claim of a sensitivity a billion times greater than that provided by his original theory would get reduced to an effective thousand — still far from sufficient to provide observability.

A detailed analysis of the sensitivity requirements for various sources according to their strengths and distances will be presented subsequently [4]. Of course, the approximate symmetry definition is only a proposal. The general prediction that the waves will be attenuated according to the number of waves that fit into the distance traversed is robust, but the explicit dependence as $1/\sqrt{n}$ is not. This is a gap in the theory that only experiments can fill. If the expected signals are *not* seen despite the design accuracy being sufficient, our proposal could be tentatively accepted till such time as the sensitivity of observation *is* adequate to test it. Of course, since we know the output of waves from the Hulse-Taylor binary pulsar, a good test would be a sufficiently accurate observation of signals from it.

Consider the Hulse-Taylor binary pulsar as the source of the waves. It has been running for a long time and the attenuation factor given should apply. However, an exploding supernova sends its waves into a space that is nearly flat. The point is that the fabric of the spacetime is being stretched by the waves but the full stretching will not occur right at the beginning. Consequently, the waves should not be attenuated that much in the beginning. After a while the space around it has been stretched as much as it can be and the waves are merely maintaining the stretching. Consequently, the attenuation should be greater later. It would be very interesting to see this effect. However, the build-up of the attenuation factor must be extremely rapid and it does not seem likely that this effect will be observable.

The proposal may seem to be so outlandish as to be dismissed out of hand. The reason is that one assumes that for gravitational waves linearized gravity would provide good guidelines and nonlinear effects would appear as perturbations on it. Indeed, that is how the theory of gravitational waves is normally developed. However, there are two points to note. For nonlinear systems there is no guarantee that one has not entered into a "chaotic" regime. The parable of the butterfly (let us call it Osama) flapping its wings at one place in the World (say Abbottabad) causing a hurricane three days later at a place nearly antipodal to it (let us say New York), is worth bearing in mind. The other point is that a "perturbation", by definition, is negligible compared to the main effect. For a zero energy, it is not clear what would be a sufficiently small change to qualify as a "perturbation".

The modern laser interferometer gravitational wave detectors came on line in June-July 2009. At the time there had even been hopes that the waves might have been detected by mid-July 2009. However, in two years of running no signal has been seen. There are now limits being put on the strengths of sources on the basis of the lack of a signal [18]. Since the proposal discussed in this paper pre-dates even the start of the observations, leave alone the lack of observations [9], it is worth considering this proposal as a contender for a valid definition of energy in GR.

It is interesting to note that Weber's claimed "quantum coherence" effect leads to much *greater* efficiency of the bar detector than the interferometer (by a factor of about a billion). As is intuitively clear, the efficiency of a longer detector should be greater. According to our proposal, the efficiency increases with the square root of the length. By this proposal the efficiency of the bar detector of 3 m length is about 30 times *less* than the interferometer of 3 km arm. We urgently need that both the bar detector and interferometer come on-line to provide the data so desperately needed.

Acknowledgments

I am grateful to F. De Paolis for very useful comments on an earlier version of this paper.

References

[1] Baikov V.A., Gazizov R.K., Ibragimov N.H., 1989, *Math. USSR Sbornik*, **64**, 427
[2] Bassan M., 1993, Statement at discussion session at the Symposium on Experimental Gravitation, Nathiagali, Pakistan, 1993
[3] Bona C., Carot J., Palenzueala-Luque C., 2005, *Phys. Rev.*, **D72**, 1240101
[4] De Paolis F., Qadir A., Work in progress
[5] Ehlers J., Kundt W., 1962, "Exact solutions of gravitational field equations", *In Gravitation: An Introductoion to Current Research*, pp.49 - 101, eds. Witten L, Wiley, New York
[6] Hulse R.A., Taylor J.H., 1975, *Astrophys. J.* **195**, L51
[7] Hussain I., Mahomed F.M., Qadir A., 2007, *SIGMA* **3**, 115
[8] Hussain I., Mahomed F.M., Qadir A., 2009a, "Approximate Noether symmetries of the geodesic equations for the charged-Kerr spacetime and rescaling of energy", *Gen. Rel. Grav.*, **41**, 2399
[9] Hussain I., Mahomed F.M., Qadir A., 2009b, "A proposal for determining the energy content of gravitational waves by using approximate symmetries of differential equations", *Phys. Rev.* **D 79**, 125014 (14 pages)
[10] Hussain I., Qadir A., 2008, *Nuovo Cimento*, **B122**, 593
[11] Ibragimov N.H., *Elementary Lie Group Analysis and Ordinary Differential Equations*, Wiley, Chichester, 1999
[12] Isaacson R.A., 1968, *Phys. Rev.*, **166**, 1263; **166**, 1272
[13] Kara A.H., Mahomed F.M., Qadir A., 2008, *Nonlinear Dyn.*, **51**, 183
[14] Khan K.A., Penrose R., 1971, *Nature: Physical Science*, **229**, 185
[15] Komar A., 1963, *Phys. Rev.*, **127** (1962) 1411; **129**, 1873
[16] Landau LD, 1946, *J. Phys.*, (USSR), **10**, 25
[17] Landau L.D., Lifshitz E.M., 1987, *The Classical Theory of Fields*, Pergamon Press, Oxford
[18] Leaci, P., 2011, "Searching for gravitational wave signals" *Third Joint Italian-Pakistani Workshop on Relativistic Astrophysics, June 21-23, 2011*, talk presented, to appear in this volume
[19] Lie S. 1883, "Klassifikation und integration von gewnlichen differentialgleichungenzwischen x, y, die eine gruppe von transformationen gestaten", *Arch Math* **VIII** (IX): 187453
[20] Matzner R., 1968*J. Math. Phys.*, **9**, 1657; 1063
[21] Misner C.W., Thorne K.S., Wheeler J.A., 1973, *Gravitation* WH Freeman, pp. 464
[22] Nutku Y., 1991, *Phys. Rev.* **D 44**, 3164
[23] Preparata G., 1990, Modern Phys. Lett. **A 5**, 1
[24] Qadir A., Sharif M., 1992, *Physics Lett.*, **A167**, 331
[25] Spero A., Baierlein R., 1977, 1978, *J. Math. Phys.*, **18**, 1330; **19**, 1324
[26] Szekeres P., 1972, *J. Math. Phys.* **13**, 286
[27] Taub A.H., *Ann. Math.*, **53** (1951) 472
[28] Wang L., Wheeler J.C., Höflich P. et al., 2002, *Astrophys. J.* **579**, 671
[29] Weber J., 1994, *Experimental Gravitation: Proceedings of the International Symposium on Experimental Gravitation*,pp 251 - 260, eds. Karim M., Qadir A., IOP Press
[30] Weber J., 1961, *General Relativity and Gravitational Waves*, Interscience, New York
[31] Weber J., 1984, 1986, *Foundations of Physics*, **14**, 1185; *Sir Arthur Eddington Centeenary Symposium Vol. 3*, Ch. 1, World Scientific
[32] Weber J., Wheeler J.A., 1957, *Rev. Mod. Phys.,* **29**, 509
[33] York J.W., 1974, *Ann. Inst. Henri Poincaré XXI,* **4**, 319.

Journal of Physics: Conference Series **354** (2012) 012015 doi:10.1088/1742-6596/354/1/012015

Thermodynamics of topological black holes in Hořava–Lifshitz gravity

Hernando Quevedo[1,2], **Alberto Sánchez**[3], **and Safia Taj**[2,4]

[1]Instituto de Ciencias Nucleares, Universidad Nacional Autónoma de México, AP 70543, México, DF 04510, Mexico

[2]Dipartimento di Fisica and ICRA, Università di Roma La Sapienza, I-00185 Roma, Italy

[3]Departamento de Posgrado, CIIDET, AP 752, Querétaro, QRO 76000, Mexico

[4]Center for Advanced Mathematics and Physics, National University of Sciences and Technology, H-12, Islamabad, Pakistan

E-mail: quevedo@nucleares.unam.mx, asanchez@nucleares.unam.mx, safiataaj@gmail.com

Abstract. We investigate the thermodynamic properties of the most general static, spherically symmetric, topological black holes of the Hořava-Lifshitz gravity. The mass of these black holes turns out to depend on the entropy and the cosmological constant that we consider as an additional thermodynamic variable. We study different thermodynamic ensembles and the corresponding heat capacities in order to show that second-order phase transitions occur at certain specific values of the horizon radius. Moreover, we study the geometric properties of the equilibrium space by using the formalism of geometrothermodynamics. We show that the phase transitions, which are characterized by divergencies of the heat capacities, are described in geometrothermodynamics by curvature singularities in the equilibrium space.

1. Introduction

Recently, a field theoretical model that can be interpreted as a complete theory of gravity in the ultraviolet (UV) limit was recently proposed by Hořava [1, 2]. The model is renormalizable [3] and non-relativistic in the UV regime. Moreover, in the infrared (IR) limit it can be reduced to Einstein's gravity theory with a cosmological constant. In this model, space and time have different scalings at the UV fixed point, i. e., $x^i \to lx^i$, $t \to l^z t$ where z is the scaling exponent. In particular, it is renormalizable if $z = 3$. It is usually named Hořava-Lifshitz (HL) theory in the literature [4, 5] to emphasize the anisotropy of space and time.

It was found that the Schwarzschild–anti de Sitter black hole solution is not recovered in the IR limit [6]. This difficulty was solved by introducing an additional parameter which modifies the IR behavior [7, 8, 9]. This theoretical model of quantum gravity has been extensively investigated in the context of black hole physics and classical cosmology (see, for instance, [10, 11, 12, 13, 14, 15, 16, 17, 18, 19] and the references cited therein).

The HL gravity breaks general 4D covariance and splits it into 3D covariance plus reparametrization invariance of time. It is therefore convenient to formulate it in the $(3 + 1)$–ADM formalism, where an arbitrary metric can be written in the form

$$ds^2 = -N^2 dt^2 + g_{ij}(dx^i + N^i dt)(dx^j + N^j dt)\,, \tag{1}$$

where N^2 is the lapse function and N^i represents the shift. Then, the HL action is written as [20],

$$\mathcal{I}_{HL} = \int \mathcal{L}_{HL}\, dt\, d^3x\,, \tag{2}$$

where

$$\mathcal{L}_{HL} = \sqrt{g}N \left[\frac{2}{\kappa}\Big(K_{ij}K^{ij} - \lambda K^2\Big) + \frac{\kappa^2\mu^2(\Lambda R - 3\Lambda^2)}{8(1-3\lambda)} + \frac{\kappa^2\mu^2(1-4\Lambda)}{32(1-3\lambda)}R^2 - \right.$$
$$\left. - \frac{\kappa^2}{2\omega^4}\left(C_{ij} - \frac{\mu\omega^2}{2}R_{ij}\right)\left(C^{ij} - \frac{\mu\omega^2}{2}R^{ij}\right)\right]. \tag{3}$$

Here R_{ij} and R are the 3D Ricci tensor and curvature scalar, respectively. Moreover, the extrinsic curvature K_{ij} and the Cotton tensor C_{ij} are given by the expressions

$$K_{ij} = \frac{1}{2N}(\dot{g}_{ij} - \nabla_i N_j - \nabla_j N_i)\,, \quad C^{ij} = \epsilon^{ikl}\nabla_k\left(R^j{}_l - \frac{1}{4}\,R\,\delta^j{}_l\right), \tag{4}$$

where a dot represents differentiation with respect to the time coordinate. Finally, κ^2, λ, μ, ω and Λ are constants parameters.

The vacuum of this theory turns out to be the anti–de Sitter spacetime; however, it is possible to consider an additional term ($\mu^4 R$) in the original action to obtain a Minkowski vacuum in the IR limit. This generalization is known as the deformed HL model.

A comparison of the HL action with the Einstein-Hilbert action leads to the conclusion that the speed of light, Newton's constant and the cosmological constant $\tilde{\Lambda}$ are given by

$$c = \frac{\kappa^2\mu}{4}\sqrt{\frac{\Lambda}{1-3\lambda}}\,, \quad G = \frac{\kappa^2 c}{32\pi}\,, \quad \tilde{\Lambda} = \frac{3}{2}\Lambda\,. \tag{5}$$

Consider now the spherically symmetric line element

$$ds^2 = -\tilde{N}^2(r)f(r)dt^2 + \frac{dr^2}{f(r)} + r^2 d\Omega_k^2\,, \tag{6}$$

where $d\Omega_k^2$ is the line element of the 2–dimensional Einstein space with constant curvature $2k$. Substituting the metric (6) into the action (2), we obtain [21],

$$\mathcal{I}_{HL} = \frac{\kappa^2\mu^2\Omega_k}{8(1-3\lambda)}\int \tilde{N}\left[\frac{(\lambda-1)}{2}F'^2 - \frac{2\lambda}{r}FF' + \frac{(2\lambda-1)}{r^2}F^2\right] dt\, dr\,, \tag{7}$$

where a prime denotes the derivative with respect to r, and F is defined as,

$$F(r) = k - \Lambda r^2 - f(r)\,. \tag{8}$$

The variation of (7) leads to the following set of equations

$$\left(\frac{2\lambda}{r}F - (\lambda-1)F'\right)\tilde{N}' + (\lambda-1)\left(\frac{2}{r^2}F - F''\right)\tilde{N} = 0\,, \tag{9}$$

$$(\lambda-1)r^2F'^2 - 4\lambda rFF' + 2(2\lambda-1)F^2 = 0\,, \tag{10}$$

whose solution is

$$F(r) = \alpha r^s , \quad \tilde{N} = \gamma r^{1-2s} , \tag{11}$$

where α and γ are integration constants and s is given by

$$s = \frac{2\lambda - \sqrt{2(3\lambda - 1)}}{\lambda - 1} . \tag{12}$$

This solution was obtained recently by Cai, Cao and Ohta (CCO) [22]. In general, the value of s with a positive sign in front of the square root is also a solution of the above equations. However, in this case the asymptotic properties of the solution are not compatible with the properties of a black hole spacetime. In the allowed interval $\lambda > 1/3$, i. e. for $s \in (-1, 2)$, the above solution is asymptotically anti–de Sitter and describes the gravitational field of a static black hole.

The main goal of the present work is to analyze the thermodynamic properties and the geometry of the equilibrium space of the CCO topological black holes. In particular, we find the different heat capacities that are derived by using different ensembles. It follows that these black holes posses a phase transition structure that is characterized by divergencies of the heat capacities, indicating that the phase transitions are of second order. This paper is organized as follows. In Sec. 2, we derive the most significant thermodynamic variables of the CCO black holes and study their behavior. The heat capacities are calculated explicitly and their divergencies are discussed. In Sec. 3, we briefly review the formalism of geometrothermodynamics that is then used to study the geometric properties of the equilibrium space. We show that the thermodynamic properties of the CCO black holes can be described correctly in terms of the geometric properties of the equilibrium space. Finally, in Sec. 4 we present the conclusions of our work. Throughout this work we use units with $c = G = \hbar = k_B = 1$.

2. Thermodynamics

The most popular method to derive the thermodynamic variables consists in computing the total energy of the system in terms of physical variables, say E^a with $a = 1, \ldots, n$. In the case of black holes, the total energy coincides with the total mass M so that $M = M(E^a)$. Moreover, usually one of the variables E^a can be associated with the area of the outermost horizon. Then, if we invoke the fundamental assumption of black hole thermodynamics, according to which the area of the horizon is proportional to the entropy, it turns out that the total mass depends on the entropy, i. e., $M = M(S, E^i)$ with $i = 2, \ldots, n$. It then follows that $M = M(S, E^i)$ represents the fundamental thermodynamic equation from which the remaining variables can be derived. In fact, if we consider M as a thermodynamic potential, and assume the validity of the first law of thermodynamics

$$dM = TdS + I_i dE^i , \qquad I_i = \frac{\partial M}{\partial E^i} , \tag{13}$$

the thermodynamic variables I_i dual to E_i can easily be computed. This strategy has been widely used in the context of black hole thermodynamics.

To determine the total mass of the CCO black hole, it is convenient to use the Hamiltonian approach that yields [21] the mass

$$M = M_0 l^{2-2s} \left[\frac{k + \frac{r_+^2}{l^2}}{\left(\frac{r_+}{l}\right)^s} \right]^2 , \quad M_0 = \frac{c^3 \gamma \Omega_k}{16\pi G} \left(\frac{1+s}{2-s} \right) . \tag{14}$$

In the interval of values allowed for the parameter s the total mass turns out to be posivite definite. Furthermore, the entropy can be shown to be given as

$$S = \frac{4\pi M_0 l^2}{\gamma} \left(\frac{r_+^2}{l^2} + k \ln \frac{r_+^2}{l^2} \right) . \tag{15}$$

Here $l^2 = -1/\Lambda$ is the curvature radius and r_+ represents the radius of the outer horizon that is a function of M and l determined by the algebraic equation

$$r_+^s - \frac{\mathbb{A}}{M^{\frac{1}{2}}l^2}r_+^2 - \frac{\mathbb{A}k}{M^{\frac{1}{2}}} = 0\,, \quad \mathbb{A} = \frac{\kappa\mu\gamma^{\frac{1}{2}}\Omega_k^{\frac{1}{2}}}{2^{\frac{7}{4}}[3\lambda - 1]^{\frac{1}{4}}}\,. \tag{16}$$

According to the above equations, the total mass depends implicitly on S and explicitly on l. Although only S is clearly a thermodynamic variable, we assume here that the curvature radius l can also be considered as a thermodynamic parameter, and we will investigate the consequences that follow from this assumption. Then, the first law of thermodynamics for this case reads

$$dM = TdS - Ldl \tag{17}$$

where L is the thermodynamic variable dual to the curvature radius l. It is then straightforward to derive the expressions for the Hawking temperature T and the dual variable L as

$$T = \frac{\gamma}{4\pi r_+^{2s}}\left[(2-s)\frac{r_+^2}{l^2} - ks\right]\,, \tag{18}$$

$$L = 2M_0 l r_+^{-2s}\left(\frac{r_+^4}{l^4} - k^2\right)\,. \tag{19}$$

The behavior of the temperature is illustrated in Fig.1 for different values of the parameters. Notice that for all the values of k there exist unphysical regions with $T < 0$ that must be excluded from the analysis.

2.1. Phase transitions

According to Davies [23, 24, 25], second order phase transitions take place at those points where the heat capacity diverges. In this context, it is important to define the "heat" for the black hole. To this end, we simply use the analogy with ordinary thermodynamics and rewrite the first law of thermodynamics as

$$dM = dQ - Ldl \tag{20}$$

where the "heat" is assumed to be determined as $dQ = TdS$. Then, keeping the value of l constant we can define

$$C_l = \left(\frac{\partial Q}{\partial T}\right)_l = \left(\frac{\partial M}{\partial T}\right)_l = \left(\frac{\partial M}{\partial r_+}\right)_l \left(\frac{\partial T}{\partial r_+}\right)_l^{-1}\,. \tag{21}$$

In the present case, using the expressions (14) and (18), we obtain

$$C_l = \frac{4\pi M_0 l^2}{\gamma}\left[\frac{\left(k + \frac{r_+^2}{l^2}\right)\left[(2-s)\frac{r_+^2}{l^2} - ks\right]}{(s-1)(s-2)\frac{r_+^2}{l^2} + ks^2}\right]\,. \tag{22}$$

It follows that the singularities of the heat capacity are located at

$$\frac{r_+^2}{l^2} = \frac{ks^2}{(s-1)(2-s)}\,. \tag{23}$$

We see that phase transitions can occur only for $k = 1$ and $s \in (1, 2)$, and for $k = -1$ and $s \in (-1, 1)$. For all the remaining values of k and s, the corresponding black hole cannot

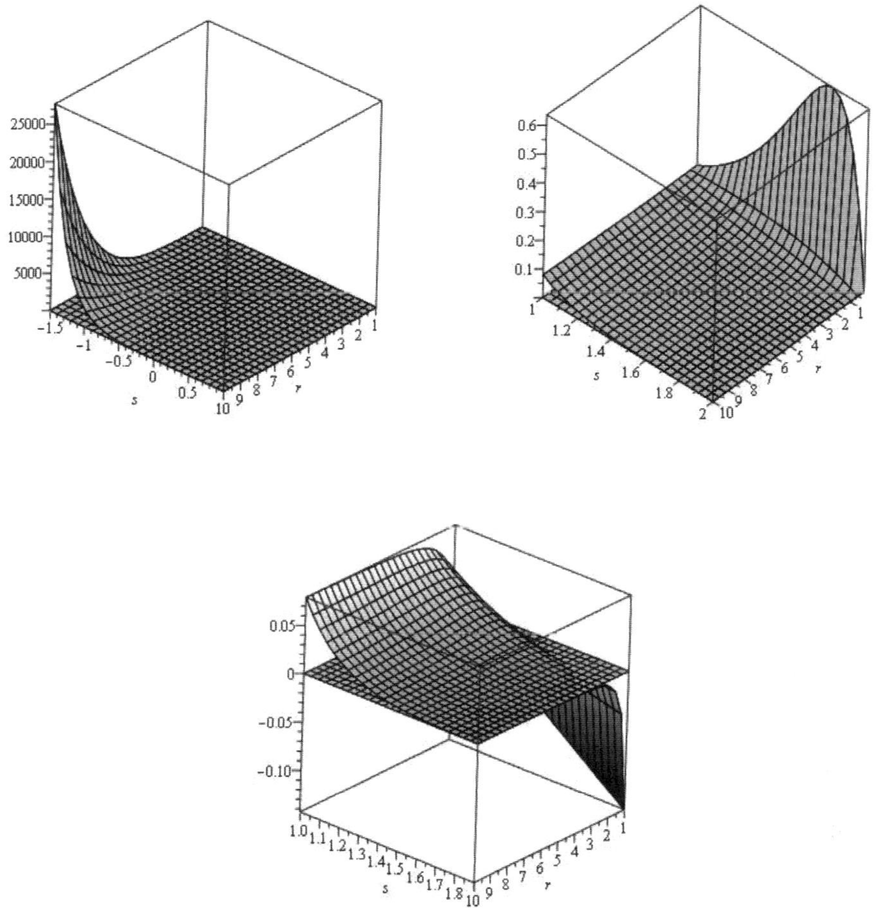

Figure 1. Behavior of the temperature for $k = -1$, $k = 0$ and $k = +1$, respectively, in terms of the horizon radius r_+ and the parameter s. Here we set $l = 1$ and $\gamma = 1$. Only the regions with $T > 0$ are physically allowed.

undergo a phase transition. Notice, however, that the phase transition condition (23) must be considered together with the inequality

$$\frac{r_+^2}{l^2} > \frac{ks}{2-s} \tag{24}$$

that follows from the condition $T > 0$ from Eq.(18).

An additional heat capacity can be obtained by applying the partial Legendre transform

$$M_1 = M + lL \tag{25}$$

that corresponds to a different thermodynamic ensemble. Then, the first law of thermodynamics reads

$$dM_1 = dQ + ldL \ . \tag{26}$$

A straightforward computation yields

$$M_1 = M_0 l^2 r^{-2s}\left(3\frac{r_+^4}{l^4} + 2k\frac{r_+^2}{l^2} - k^2\right) . \tag{27}$$

It is then clear that for this ensemble we can define an additional heat capacity that takes into account the change of "heat" in terms of the temperature for a fixed L, i. e.,

$$C_L = \left(\frac{\partial Q}{\partial T}\right)_L = \left(\frac{\partial M_1}{\partial T}\right)_L = \frac{4\pi M_0 l^2}{\gamma}\left[\frac{(2s-3)\frac{r_+^4}{l^4} + 2k\frac{r_+^2}{l^2} - 3k^2}{(s-1)(s-2)\frac{r_+^2}{l^2} + ks^2}\right] . \tag{28}$$

We see that the divergencies of C_L coincide with the roots of the equation (23); this means that both C_L and C_l show the same phase transition structure for the CCO black hole.

3. Geometrothermodynamics

Geometrothermodynamics (GTD) is a theory that has been formulated recently [26] in order to introduce in a consistent manner the Legendre invariance in the geometric description of the space of thermodynamic equilibrium states. In all the cases analyzed so far, GTD has delivered consistent results and allows us to describe geometrically the thermodynamic interaction and the phase transitions by means of Legendre invariant metrics. This theory has been applied to different thermodynamic systems the ideal gas, the van der Waals gas and a diverse black holes in different theories [27, 28, 29, 30, 31, 32]. In this Section we will show that GTD correctly describes the thermodynamic properties of the CCO black holes.

The main ingredient of GTD is a $(2n+1)$–dimensional manifold $\mathcal{T}$ where n is the number of thermodynamic degrees of freedom of the system. We introduce in $\mathcal{T}$ the set of coordinates $Z^A = (\Phi, E^a, I^a)$, where Φ is an arbitrary thermodynamic potential, E^a, $a = 1, 2, ..., n$, are the extensive variables, and I^a the intensive variables. It is also possible to introduce in a canonical manner the fundamental one–form $\Theta = d\Phi - \delta_{ab} I^a dE^b$, $\delta_{ab} = \text{diag}(+1, ..., +1)$, that satisfies the condition $\Theta \wedge (d\Theta)^n \neq 0$, and is invariant with respect to Legendre transformations $(\Phi, E^a, I^a) \to (\tilde\Phi, \tilde E^a, \tilde I^a)$ with $\Phi = \tilde\Phi - \delta_{ab}\tilde E^a \tilde I^b$, $E^a = -\tilde I^a$, and $I^a = \tilde E^a$. Moreover, we assume that on $\mathcal{T}$ there exists a metric G which is also invariant with respect to Legendre transformations. The triad $(\mathcal{T}, \Theta, G)$ defines a Riemannian contact manifold that is called the thermodynamic phase space (phase manifold). The space of thermodynamic equilibrium states (equilibrium manifold) is an n–dimensional Riemannian submanifold $\mathcal{E} \subset \mathcal{T}$ induced by a smooth map $\varphi : \mathcal{E} \to \mathcal{T}$, i.e. $\varphi : (E^a) \mapsto (\Phi, E^a, I^a)$, with $\Phi = \Phi(E^a)$ and $I^a = I^a(E^a)$, such that $\varphi^*(\Theta) = \varphi^*(d\Phi - \delta_{ab} I^a dE^b) = 0$ holds, where φ^* is the pullback of φ. The manifold $\mathcal{E}$ is naturally equipped with the Riemannian metric $g = \varphi^*(G)$. The purpose of GTD is to demonstrate that the geometric properties of $\mathcal{E}$ are related to the thermodynamic properties of a system with fundamental thermodynamic equation $\Phi = \Phi(E^a)$.

The nondegenerate Legendre invariant metric [33]

$$G = \Theta^2 + \frac{1}{2}\left[(\delta_{ab} - \eta_{ab})E^a I^b\right]\left(\eta_{cd}\, dE^c dI^d\right) , \tag{29}$$

where $\eta_{ab} = \text{diag}(-1, 1, \dots, 1)$, has been used extensively to describe second order phase transitions, especially in the context of black hole thermodynamics. The metric (29) induces on $\mathcal{E}$, by means of $g = \varphi^*(G)$, the thermodynamic metric

$$g = \frac{1}{2}\left[E^a\left(\frac{\partial\Phi}{\partial E^a} - \eta_{ab}\delta^{bc}\frac{\partial\Phi}{\partial E^c}\right)\right]\left(\eta_{ab}\delta^{bc}\frac{\partial^2\Phi}{\partial E^c \partial E^d} dE^a dE^d\right) . \tag{30}$$

Notice that the metric g can be computed explicitly once the fundamental equation $\Phi = \Phi(E^a)$ is given whose existence is assumed to be guaranteed by the definition of the smooth map φ. Moreover, the second law of thermodynamics is equivalent to the condition

$$\pm \frac{\partial^2 \Phi}{\partial E^a \partial E^b} \geq 0 \ , \tag{31}$$

where the sign $\pm$ depends on the choice of the thermodynamic potential. For instance, the positive sign must be selected, if Φ is chosen as the entropy. In the energy representation, Φ corresponds to the internal energy and the sign must be negative.

For the CCO topological black hole let us take, for instance, $\Phi = M$. The coordinates of the 5-dimensional phase manifold can be chosen as $Z^A = (M, S, l, T, L)$, where T is the temperature dual to S and L is the dual of the curvature radius l. The fundamental one-form is then $\Theta = dM - TdS + Ldl$ and the Legendre invariant metric (29) is written as

$$G = \Theta^2 + ST\left(-dSdT + dldL\right) \ . \tag{32}$$

The smooth map $\varphi : \mathcal{E} \rightarrow \mathcal{T}$ or in coordinates

$$\varphi : (S, l) \mapsto \{M(S,l), S, l, T(S,l), L(S,l)\} \tag{33}$$

determines the equilibrium manifold $\mathcal{E}$ with metric

$$g = \varphi^*(G) = SM_S\left(-M_{SS}dS^2 + M_{ll}dl^2\right) \ , \tag{34}$$

on which the first law of thermodynamics $dM = TdS - Ldl$ and the equilibrium conditions

$$T = \frac{\partial M}{\partial S} \equiv M_S \ , \quad -L = \frac{\partial M}{\partial l} \equiv M_l \tag{35}$$

hold.

In this particular case, the fundamental equation $M = M(S,l)$ cannot be written explicitly and so we use r_+ instead of S as a coordinate. Then,

$$g = SM_S\left[-M_{SS}S_{r_+}^2 dr_+^2 - 2M_{SS}S_{r_+}S_l dl dr_+ + (M_{ll} - M_{SS}S_l^2)dl^2\right] \ . \tag{36}$$

Using the expressions for the mass and the entropy, we obtain

$$\begin{aligned} g \ &= \ \frac{c^3\Omega_k^2\gamma^2 r_+^{-4s}}{16\pi^2 G}\left(\frac{s+1}{2-s}\right)\left[(s-2)\frac{r_+^2}{l^2} + ks\right] S(r_+,l)\Bigg\{\left[(s-1)(s-2)\frac{r_+^2}{l^2} + ks^2\right] \\ &\times \left[\left(1 + \frac{kl^2}{r_+^2}\right)dr_+^2 + 2\frac{kl}{r_+}\left(\ln\frac{r_+^2}{l^2} - 1\right)dl dr_+\right] + F(r_+,l)dl^2\Bigg\} \ , \end{aligned} \tag{37}$$

where

$$F(r_+,l) = \frac{2k^2\left(\ln\frac{r_+^2}{l^2} - 1\right)^2\left[(s-1)(s-2)\frac{r_+^2}{l^2} + ks^2\right] - \left(k^2 + 3\frac{r_+^4}{l^4}\right)\left(k + \frac{r_+^2}{l^2}\right)}{2\left(k + \frac{r_+^2}{l^2}\right)}. \tag{38}$$

Journal of Physics: Conference Series **354** (2012) 012015 doi:10.1088/1742-6596/354/1/012015

The curvature scalar corresponding to the metric (36) can be computed explicitly as

$$R = \frac{N}{D}\,, \quad D = \left(k^2 + 3\frac{r_+^4}{l^4}\right)^2 \left(k + \frac{r_+^2}{l^2}\right)^3 \left(\frac{r_+^2}{l^2} + k\ln\frac{r_+^2}{l^2}\right)^3 \times \left[(2-s)\frac{r_+^2}{l^2} - ks\right]^4 \left[(s-1)(s-2)\frac{r_+^2}{l^2} + ks^2\right]^2 \tag{39}$$

where N is a function of r_+ and l that is finite at those points where the denominator vanishes. Notice that there are several curvature singularities. The first set of singularities is situated at the roots of $k + r_+^2/l^2 = 0$. As follows from Eq.(14), these singularities correspond to the limit $M \to 0$. A second singularity is located at the roots of the equation $r_+^2 + k\ln r_+^2 = 0$; from Eq.(15) it follows that it can be interpreted as the limit in which the entropy vanishes. Furthermore, according to Eq.(18), the singularity situated at $r_+^2/l^2 = ks/(2-s)$ corresponds the limit $T \to 0$. Finally, if $(s-1)(s-2)r_+^2/l^2 + ks^2 = 0$ a singularity occurs that, according to Eq.(22), coincides with the limit $C \to \infty$, i. e., with the points where second order phase transitions take place. The singularities at which the mass, the entropy or the temperature vanish indicate the limit of applicability of the thermodynamics of black holes and of GTD. We conclude that the curvature obtained within the formalism of GTD describes correctly and in an invariant manner the thermodynamic behavior of topological black holes in HL gravity.

4. Conclusions

In this work, we studied the thermodynamic properties of the Cai-Cao-Ohta (CCO) topological black holes in the Hořava-Lifshitz model of quantum gravity. We assumed the validity of the first law of thermodynamics of black holes to show the consistency of the thermodynamic variables. In particular, we derived two different heat capacities C_l and C_L that turned out to predict the existence of a non-trivial phase transition structure. In fact, it was shown that both heat capacities diverge at certain values of the horizon radius, indicating that second-order phase transitions occur at the singular points.

We applied the formalism of geometrothermodynamics (GTD) to describe the thermodynamics of the CCO black holes in terms of geometric concepts. In the thermodynamic phase manifold we introduce a particular Riemannian metric whose main property is its invariance with respect to Legendre transformations, i. e., its geometric characteristics are independent of the choice of thermodynamic potential. This is a property that holds in ordinary thermodynamics and that we assume as valid in GTD too. The Legendre invariant metric induces in a canonical manner a thermodynamic metric in the equilibrium manifold that is defined as a submanifold of the thermodynamic phase manifold.

We used the expressions for the main thermodynamic variables of the CCO black holes in order to compute the explicit form of the thermodynamic metric of the equilibrium manifold. The corresponding thermodynamic curvature turned out to be nonzero in general, indicating the presence of thermodynamic interaction. Moreover, it was shown that the phase transitions, which are characterized by divergencies of the heat capacities, are described in GTD by curvature singularities in the equilibrium manifold.

It was found that the geometrothermodynamic equilibrium manifold of the CCO black holes present additional curvature singularities which correspond to the vanishing of the mass, entropy and Hawking temperature. We interpret in general the vanishing of these thermodynamic variables as an indication of the limit of applicability of black hole thermodynamics. So we conclude that the formalism of GTD breaks down, with curvature singularities, exactly at those points where black hole thermodynamics fails.

Acknowledgments
Two of us (HQ and ST) would like to thank the ICRANet for support and hospitality, and Professor Asghar Qadir for encouragement and support. This work was supported in part by DGAPA-UNAM, grant No. IN106110.

References

[1] Hořava P 2009 *Phys. Rev. Lett.* **102** 161301
[2] Hořava P. 2009 *Phys. Rev.* **D79** 084008 [hep–th/0901.3775]
[3] Orlando D and Reffert S 2009 *Class. Quant. Grav.* **26** 155021 [hep–th/ 0905.0301]
[4] Boehmer C G and Lobo F S N 2010 *Eur. Phys. J.* **C70** 1111 [gr–qc/0909.3986]
[5] Blas D, Pujolas O and Sibiryakov S 2010 **104** 181302 [hep–th/0909.3525]
[6] Park M I 2009 *J. High Energy Phys.* **0909** 123 [hep–th/0905.4480]
[7] Lu H, Mei J and Pope C N 2009 *Phys. Rev. Lett.* **103** 091301 [hep–th/0904.1595]
[8] Nastase H *On IR solutions in Horava gravity theories*, [hep–th/0904.3604]
[9] Cai R G and Ohta N 2010 Phys. Rev. **D81** 084061[hep–th/0910.2307]
[10] Koutsoumbas G, Papantonopoulos E, Pasipoularides P and Tsoukalas M *Black hole solutions in 5D Horava-Lifshitz gravity*, [hep–th/1004.2289]
[11] Koutsoumbas G and Pasipoularides P *Black hole solutions in Horava- Lifshitz gravity with cubic terms*, [hep–th/1006.3199].
[12] Majhi B R 2010 *Phys. Lett.* **B686** 49 [hep–th/0911.3239]
[13] Saridakis E N 2010 *Eur. Phys. J.* **C67** 229 [hep–th/0905.3532]
[14] Bogdanos C and Saridakis E N 2010 *Class. Quant. Grav.* **27** 075005 [hep–th/0907.1636]
[15] Leon G and Saridakis E N 2009 *JCAP* **0911** 006 [hep–th/0909.3571]
[16] Dutta S and Saridakis E N *Observational constraints on Horava–Lifshitz cosmology*, [hep–th/0911.1435]
[17] Dutta S and Saridakis E N 2010 *JCAP* **1005** 013 [hep–th/1002.3373]
[18] Jamil M and Saridakis E N 2010 *JCAP* **1007** 028 [hep–th/1003.5637]
[19] Ali A, Dutta S, Saridakis E N and Anjan A Sen *Horava–Lifshitz cosmology with generalized Chaplygin gas*, [astro–ph/1004.2474]
[20] Lu H, Mei J and Pope C N 2009 *Phys. Rev. Lett.* **103** 091301
[21] Cai R G, Cao L M and N Ohta 2009 *Phys. Lett.* **B679** 504 [hep–th/0905.0751]
[22] Cai R G, Cao L M and Ohta N 2009 *Phys. Rev.* **D80** 024003 [hep–th/0904.3670]
[23] Davies P C W 1977 *Proc. Roy. Soc. Lond. A* **353** 499
[24] Davies P C W 1977 *Rep. Prog. Phys.* **41** 1313.
[25] Davies P C W 1989 *Class. Quant. Grav.* **6** 1909
[26] Quevedo H 2007 *J. Math. Phys.* **48** 013506
[27] Álvarez J L, Quevedo H and Sánchez A 2008 *Phys. Rev.* **D77** 084004 [gr–qc/0801.2279]
[28] Quevedo H and Sánchez A 2008 *J. High Energy Phys.* **09** 034
[29] Vázquez A, Quevedo H and Sánchez A 2010 *J. Geom. Phys.* **60**, 1942
[30] Quevedo H, Sánchez A and Vázquez A 2008 *Invariant geometry of the ideal gas* [math-ph/0811.0222]
[31] Quevedo H and Sánchez A 2009 *Phys. Rev.* **D79** 024012
[32] Quevedo H and Sánchez A 2009 *Phys. Rev.* **D79** 087504
[33] Quevedo H, Sánchez A, Taj S and Vázquez A 2011 *Gen. Rel. Grav.* **43** 1153
[34] Janke W, Johnston D A and Kenna R 2010 *J. Phys.* **A43** 425206 [hep–th/1005.3392]

The Plebański-Demiański class of black holes and quantum tunneling

K. Saifullah

Department of Mathematics, Quaid-i-Azam University, Islamabad, Pakistan

E-mail: saifullah@qau.edu.pk

Abstract. The Plebański-Demiański solution of Einstein's field equations represents a general class of black hole solutions. The thermodynamic quantities and other results for this spacetime reduce to those for the Kerr-Newman, Reissner-Nordström, and Schwarzschild black holes. Hawking radiations of charged and uncharged fermions and scalar particles have been studied for this class of black holes using the semi-classical approach of quantum tunneling. Working out the tunneling probability of these particles across the event horizons using complex path integration yields Hawking temperature as well.

1. The Plebański-Demiański solution

The Plebański-Demiański class of solutions [1] of Einstein's equations includes the famous Kerr-Newman rotating black hole, the Kerr-NUT, the Reissner-Nordström, and the Schwarzschild black holes as special cases. This also contains the particular solution for accelerating black holes [2, 3, 4, 5]. The general form of the metric contains seven parameters which characterize the mass M, electric and magnetic charges e and g respectively, Kerr-like rotation parameter a which is equal to angular momentum per unit mass i.e. $a = J/M$, the NUT (Newman-Unti-Tamburino) parameter l, acceleration of the source α and the cosmological constant Λ. We write the general Plebański-Demiański metric in the following notation [2, 3]

$$\begin{aligned} ds^2 &= \frac{1}{\bar{\Omega}^2}\{-\frac{Q}{\rho^2}[dt - (a\sin^2\theta + 4l\sin^2\frac{\theta}{2})d\phi]^2 + \frac{\rho^2}{Q}dr^2 + \frac{\rho^2}{P}d\theta^2 \\ &\quad + \frac{P}{\rho^2}\sin^2\theta[adt - (r^2 + (a+l)^2)d\phi]^2\}, \end{aligned} \tag{1}$$

where

$$\begin{aligned} \bar{\Omega} &= 1 - \frac{\alpha}{\omega}(l + a\cos\theta)r, \\ \rho^2 &= r^2 + (l + a\cos\theta)^2, \\ P &= 1 - a_3\cos\theta - a_4\cos^2\theta, \\ Q &= (\omega^2 k + e^2 + g^2) - 2Mr + \epsilon r^2 - 2\alpha\frac{n}{\omega}r^3 - (\alpha^2 k + \frac{\Lambda}{3})r^4, \\ a_3 &= 2\alpha\frac{a}{\omega}M - 4\alpha^2\frac{al}{\omega^2}(\omega^2 k + e^2 + g^2) - 4\frac{\Lambda}{3}al, \end{aligned}$$

$$\begin{aligned} a_4 &= -\alpha^2\frac{a^2}{\omega^2}(\omega^2k+e^2+g^2)-\frac{\Lambda}{3}a^2, \\ \epsilon &= \frac{\omega^2k}{a^2-l^2}+4\alpha\frac{l}{\omega}M-(a^2+3l^2)[\frac{\alpha^2}{\omega^2}(\omega^2k+e^2+g^2)+\frac{\Lambda}{3}], \quad (2) \\ n &= \frac{\omega^2kl}{a^2-l^2}-\alpha\frac{a^2-l^2}{\omega}M+(a^2-l^2)l[\frac{\alpha^2}{\omega^2}(\omega^2k+e^2+g^2)+\frac{\Lambda}{3}], \quad (3) \\ k &= [1+2\alpha\frac{l}{\omega}M-3\alpha^2\frac{l^2}{\omega^2}(e^2+g^2)-l^2\Lambda](\frac{\omega^2}{a^2-l^2}+3\alpha^2l^2)^{-1}. \quad (4) \end{aligned}$$

Here ϵ and k are arbitrary real parameters, n is the Plebański-Demiański parameter and ω is the twist. When α, l, g and Λ vanish the line element reduces to the Kerr-Newman solution. If the electric charge and rotation parameter also vanish, we get the Schwarzschild solution. Thus the metric represents a complete class of accelerating, rotating and charged black holes. We note that the metric is singularity free if $|a| < |l|$, and it has a Kerr-like ring singularity at $\rho = 0$ when $|a| \geq |l|$. We will see later that the following form of the above metric is very convenient for many purposes [6, 7, 8, 9]

$$ds^2 = -f(r,\theta)\,dt^2 + \frac{dr^2}{g(r,\theta)} + \Sigma(r,\theta)\,d\theta^2 + K(r,\theta)\,d\phi^2 - 2H(r,\theta)\,dtd\phi, \quad (5)$$

where the functions f, g, Σ, K and H have the following definitions

$$\begin{aligned} f(r,\theta) &= \left(\frac{Q-a^2P\sin^2\theta}{\rho^2\Omega^2}\right), \quad (6) \\ g(r,\theta) &= \frac{Q\Omega^2}{\rho^2}, \quad (7) \\ \Sigma(r,\theta) &= \left(\frac{\rho^2}{P\Omega^2}\right), \quad (8) \\ K(r,\theta) &= \left(\frac{\sin^2\theta\left[P\left(r^2+a^2\right)^2-a^2Q\sin^2\theta\right]}{\rho^2\Omega^2}\right), \quad (9) \\ H(r,\theta) &= \left(\frac{a\sin^2\theta\left[P\left(r^2+a^2\right)-Q\right]}{\rho^2\Omega^2}\right). \quad (10) \end{aligned}$$

The electromagnetic vector potential is given by [10]

$$A = \frac{-er\left[dt-a\sin^2\theta d\phi\right]-g\cos\theta\left[adt-\left(r^2+a^2\right)d\phi\right]}{r^2+a^2\cos^2\theta}. \quad (11)$$

The event horizons are obtained by putting $g(r,\theta) = 0$, yielding

$$r = \frac{1}{\alpha\cos\theta}, r = \pm\frac{1}{\alpha}, \text{ and } r_\pm = M \pm \sqrt{M^2-a^2-e^2-g^2}, \quad (12)$$

where $r_\pm$ represent the outer and inner horizons similar to the Kerr-Newman black hole. The other horizons are acceleration horizons. We restrict ourselves only to non-extremal black holes i.e. when the quantity under the radical sign in Eq. (12) is positive. The angular velocity at the black hole horizon [6, 7]

$$\Omega_h = \frac{H(r_+,\theta)}{K(r_+,\theta)}, \quad (13)$$

becomes

$$\Omega_h = \frac{a}{(r_+^2 + a^2)}. \tag{14}$$

We define another function which will be needed later

$$F(r,\theta) = f(r,\theta) + \frac{H^2(r,\theta)}{K(r,\theta)}.$$

Using the values of f, K and H this takes the form

$$F(r,\theta) = \frac{PQ\rho^2}{\Omega^2\left[P(r^2+a^2)^2 - a^2 Q\sin^2\theta\right]}. \tag{15}$$

In this article we review some of the work done on this interesting class of solutions, particularly with reference to their thermodynamical properties and Hawking radiations. Results on Hawking radiations for charged and uncharged fermions and scalar particles are given. The Hamilton-Jacobi method of quantum tunneling and WKB approximation have been used to solve the Dirac and the Klein-Gordon equations. The tunneling probabilities of particles for crossing the event horizons have been worked out by complex path integration. We note that all the results for the general case reduce to the particular black holes, mentioned above, in appropriate limits.

2. Thermodynamics

In order to discuss thermodynamics of the class of black holes we have described in the first section, we first consider the special case of non-accelerating black holes. For this we put $\alpha = 0 = \Lambda$ in metric (1), so that $\omega^2 k = a^2 - l^2$ and therefore $\epsilon = 1$, $n = l$, $P = 1$, and we get [3]

$$ds^2 = \frac{Q}{\rho^2}[dt - (a\sin^2\theta + 4l\sin^2\frac{\theta}{2})d\phi]^2 - \frac{\rho^2}{Q}dr^2 - \rho^2 d\theta^2$$
$$-\frac{\sin^2\theta}{\rho^2}[adt - (r^2 + (l+a)^2)d\phi]^2, \tag{16}$$

where

$$\rho^2 = r^2 + (l + a\cos\theta)^2, Q = (a^2 - l^2 + e^2 + g^2) - 2Mr + r^2, \tag{17}$$

which is, in fact, the Kerr-Newman-NUT solution. Putting $Q = 0$ gives the location of inner and outer horizons of the black hole as [3]

$$r_\pm = M \pm \sqrt{M^2 + l^2 - a^2 - e^2 - g^2}, \tag{18}$$

where $M^2 \geq a^2 + e^2 + g^2 - l^2$.

The ergosphere of this black hole is obtained by

$$g_{tt} = 0. \tag{19}$$

$$r_+ \leq r(\theta) \leq r_a, \tag{20}$$

where r_a is the outer horizon of the corresponding Reissner-Nordström black hole with magnetic and NUT charges g and l respectively.

We find the surface gravity for this black hole by using the angular velocity [11]

$$\Omega = \frac{d\phi}{dt} = -\frac{g_{t\phi}}{g_{\phi\phi}}. \tag{21}$$

Thus we get

$$\Omega_h = \frac{a}{r_+^2 + (l+a)^2}. \tag{22}$$

or

$$\Omega_h = \frac{a}{2M^2 + 2l^2 + 2al - e^2 - g^2 + 2M\sqrt{M^2 + l^2 - a^2 - e^2 - g^2}}, \tag{23}$$

and, therefore, the surface gravity becomes [12]

$$\kappa_h = \frac{1}{2a}\Omega_h \left.\frac{dQ}{dr}\right|_{r=r_+}. \tag{24}$$

The Hawking temperature $T = \kappa_h/2\pi$ of the black hole becomes

$$T = \frac{1}{2\pi}\left[\frac{\sqrt{M^2 + l^2 - a^2 - e^2 - g^2}}{2M^2 + 2l^2 + 2al - e^2 - g^2 + 2M\sqrt{M^2 + l^2 - a^2 - e^2 - g^2}}\right], \tag{25}$$

When l and g vanish we get the temperature for the Kerr-Newman black hole.

Next, we consider accelerating black holes which are free of NUT parameter, l. Putting $l = 0,\ k = 1$, in Eq. (1) we note that $\omega = a, a_3 = 2\alpha M,\ a_4 = -\alpha^2(a^2 + e^2 + g^2) - \Lambda a^2/3$. This metric contains six arbitrary parameters $M,\ e,\ g,\ \alpha,\ a$ and Λ. In the following work we put the cosmological constant equal to zero. We see that these black holes have the inner and outer horizons identical to those of the Kerr-Newman black hole

$$r_\pm = M \pm \sqrt{M^2 - a^2 - e^2 - g^2}, \tag{26}$$

and acceleration horizons at $r = 1/\alpha$ and $r = 1/\alpha\cos\theta$.

The surface gravity of accelerating and rotating black holes is given by [13]

$$\kappa_h = \frac{\bar{\Omega}^2}{2\rho^2}\left[\frac{d\bar{Q}}{dr}\right]. \tag{27}$$

As $\bar{\Omega} \neq 0$ at $r = r_+$ we get

$$\kappa_h = \frac{\bar{\Omega}^2}{2\rho^2}2[(r-M)(1-\alpha^2 r^2) - \alpha^2 r(r^2 - 2Mr + a^2 + e^2 + g^2)]. \tag{28}$$

Since at the outer horizon, $r^2 - 2Mr + a^2 + e^2 + g^2 = 0$, therefore, putting the value of ρ^2 and $\bar{\Omega}$, we get

$$\kappa_h = \frac{(r_+ - M)}{(r_+^2 + a^2)}(1 - \alpha r_+)^3(1 + \alpha r_+). \tag{29}$$

Using Eq. (26) this can be written as explicitly as

$$\kappa_h = \frac{[1 - \alpha(M + \sqrt{M^2 - a^2 - e^2 - g^2})]^3[1 + \alpha(M + \sqrt{M^2 - a^2 - e^2 - g^2})]}{(\sqrt{M^2 - a^2 - e^2 - g^2})^{-1}[2M^2 - e^2 - g^2 + 2M\sqrt{M^2 - a^2 - e^2 - g^2}]}. \tag{30}$$

We see that from Eq. (29) the surface gravity will vanish at the acceleration horizon, $r = 1/\alpha$. If l and g vanish, this gives the surface gravity for the Kerr-Newman black hole. The relation

(30) can also be written in terms of the inner and outer horizons. For this we use the relation $r_+ - r_- = (r_+ - M) - (r_- - M) = 2(r_+ - M)$ in Eq. (29) to get [13]

$$\kappa_h = \frac{r_+ - r_-}{2(r_+^2 + a^2)}(1 - \alpha r_+)^3(1 + \alpha r_+). \tag{31}$$

In this case the angular velocity from Eq. (21) becomes

$$\Omega = \frac{a[\bar{Q} - P(r^2 + a^2)]}{\bar{Q}a^2 \sin^2\theta - P(r^2 + a^2)^2}, \tag{32}$$

which at the horizon takes the form

$$\Omega_h = \frac{a}{2M^2 - e^2 - g^2 + 2M\sqrt{M^2 - a^2 - e^2 - g^2}}. \tag{33}$$

We find that the acceleration parameter, α, is only permitted in the following range [13]

$$\alpha < \frac{1}{M + \sqrt{M^2 - a^2 - e^2 - g^2}}. \tag{34}$$

3. Hawking radiation of Dirac particles

Soon after the discovery of thermal radiations from black holes by Hawking [14, 15], different techniques were adopted to study their mathematical and physical properties. This, in fact, gave rise to a new area of investigations which lies at the interface of general relativity, quantum mechanics and thermodynamics. The methods used to study these radiations include the null geodesic method, Hamilton-Jacobi tunneling method [16, 17, 18, 19], the method of dimensional reduction [20, 21] and the so-called anomaly technique [22, 23]. The main idea is that a pair of particles is created near the event horizon of the black hole and from this pair the negative energy particle tunnels inwards. This decreases the mass of the black hole. The positive energy particle tunnels out and constitutes Hawking radiation.

Hawking radiation from the Plebański-Demiański class has been studied [7, 8, 9] using the Hamilton-Jacobi method with particular reference to accelerating and rotating black holes. In this section we briefly review the tunneling method for Dirac particles from these black holes and present results. The main idea is to solve the Dirac equation in the background of the spacetime in Eq. (5). We know that in covariant form the Dirac equation with mass of fermions, m, can be written as

$$\iota\gamma^\mu(D_\mu)\Psi + \frac{m}{\hbar}\Psi = 0, \tag{35}$$

where

$$D_\mu = \partial_\mu + \Omega_\mu, \tag{36}$$

$$\Omega_\mu = \frac{-1}{8}\Gamma_\mu^{\alpha\beta}[\gamma^\alpha, \gamma^\beta], \tag{37}$$

and $[\gamma^\alpha, \gamma^\beta]$ satisfy the commutation relations

$$[\gamma^\alpha, \gamma^\beta] = -[\gamma^\beta, \gamma^\alpha] \quad \text{if} \quad \alpha \neq \beta, \quad [\gamma^\alpha, \gamma^\beta] = 0 \quad \text{if} \quad \alpha = \beta. \tag{38}$$

We choose the following form of γ^μ

$$\gamma^t = \sqrt{\frac{(P(r^2 + a^2)^2 - Qa^2\sin^2\theta)\Omega^2}{PQ\rho^2}}\gamma^0, \gamma^r = \sqrt{\frac{Q\Omega^2}{\rho^2}}\gamma^3, \gamma^\theta = \sqrt{\frac{P\Omega^2}{\rho^2}}\gamma^1, \tag{39}$$

$$\gamma^{\phi} = \frac{\rho\Omega\gamma^2}{\sin\theta\sqrt{P(r^2+a^2)^2 - Qa^2\sin^2\theta}} + \frac{a(P(r^2+a^2)-Q)\gamma^0}{\sqrt{F(r,\theta)}(P(r^2+a^2)^2 - Qa^2\sin^2\theta)}, \tag{40}$$

where

$$\gamma^0 = \begin{pmatrix} 0 & I \\ -I & 0 \end{pmatrix}, \gamma^1 = \begin{pmatrix} 0 & \sigma^1 \\ \sigma^1 & 0 \end{pmatrix}, \tag{41}$$

$$\gamma^2 = \begin{pmatrix} 0 & \sigma^2 \\ \sigma^2 & 0 \end{pmatrix}, \gamma^3 = \begin{pmatrix} 0 & \sigma^3 \\ \sigma^3 & 0 \end{pmatrix}, \tag{42}$$

and the Pauli sigma matrices are

$$\sigma^1 = \begin{pmatrix} 0 & 1 \\ 1 & 0 \end{pmatrix}, \sigma^2 = \begin{pmatrix} 0 & -\iota \\ \iota & 0 \end{pmatrix}, \sigma^3 = \begin{pmatrix} 1 & 0 \\ 0 & -1 \end{pmatrix}. \tag{43}$$

To solve the Dirac equation in this set up we take an *ansatz* of the following form for the spin-up and spin-down solutions

$$\Psi_{\uparrow}(t,r,\theta,\phi) = \begin{pmatrix} A(t,r,\theta,\phi)\xi_{\uparrow} \\ B(t,r,\theta,\phi)\xi_{\uparrow} \end{pmatrix} \exp[\frac{\iota I_{\uparrow}(t,r,\theta,\phi)}{\hbar}], \tag{44}$$

$$\Psi_{\downarrow}(t,r,\theta,\phi) = \begin{pmatrix} C(t,r,\theta,\phi)\xi_{\downarrow} \\ D(t,r,\theta,\phi)\xi_{\downarrow} \end{pmatrix} \exp[\frac{\iota I_{\downarrow}(t,r,\theta,\phi)}{\hbar}]. \tag{45}$$

Here A, B, C, D are arbitrary functions of the coordinates, $\xi_{\uparrow}$ and $\xi_{\downarrow}$ are the eigenvectors of σ^i, and $I_{\uparrow}$ and $I_{\downarrow}$ are the corresponding actions. On using Eq. (38) the Dirac equation takes the form

$$(\iota\gamma^t\partial_t + \iota\gamma^r\partial_r + \iota\gamma^{\theta}\partial_{\theta} + \iota\gamma^{\phi}\partial_{\phi})\Psi + \frac{m}{\hbar}\Psi = 0. \tag{46}$$

We substitute the above *ansatz* into the Dirac equation and compute it term by term. Dividing by the exponential term and neglecting the terms with $\hbar$ we obtain the following four equations

$$\begin{aligned} 0 &= -B[\frac{1}{\sqrt{F(r,\theta)}}\partial_t I_{\uparrow} + \sqrt{\frac{\Omega^2 Q}{\rho^2}}\partial_r I_{\uparrow} \\ &+ \frac{a(P(r^2+a^2)-Q)}{\sqrt{F(r,\theta)}(P(r^2+a^2)^2 - Qa^2\sin^2\theta)}\partial_{\phi} I_{\uparrow}] + Am, \end{aligned} \tag{47}$$

$$0 = -B[\sqrt{\frac{\Omega^2 P}{\rho^2}}\partial_{\theta} I_{\uparrow} + \frac{\iota\rho\Omega}{\sin\theta(\sqrt{P(r^2+a^2)^2 - Qa^2\sin^2\theta)}}\partial_{\phi} I_{\uparrow}], \tag{48}$$

$$\begin{aligned} 0 &= A[\frac{1}{\sqrt{F(r,\theta)}}\partial_t I_{\uparrow} - \sqrt{\frac{\Omega^2 Q}{\rho^2}}\partial_r I_{\uparrow} \\ &+ \frac{a(P(r^2+a^2)-Q)}{\sqrt{F(r,\theta)}(P(r^2+a^2)^2 - Qa^2\sin^2\theta)}\partial_{\phi} I_{\uparrow}] + Bm, \end{aligned} \tag{49}$$

$$0 = A[\sqrt{\frac{\Omega^2 P}{\rho^2}}\partial_{\theta} I_{\uparrow} + \frac{\iota\rho\Omega}{\sin\theta\sqrt{P(r^2+a^2)^2 - Qa^2\sin^2\theta}}\partial_{\phi} I_{\uparrow}]. \tag{50}$$

Considering the Killing vectors of the background spacetime we take the following separation of variables for the action [7]

$$I_{\uparrow} = -Et + J\phi + W(r,\theta). \tag{51}$$

Here E and J are the energy and angular momentum of the emitted particle. Substituting this into the above set of four equations yields

$$\begin{aligned}0 &= -B[\frac{-E}{\sqrt{F(r,\theta)}} + \sqrt{\frac{\Omega^2 Q}{\rho^2}} W'(r,\theta) \\ &+ \frac{a(P(r^2+a^2)-Q)}{\sqrt{F(r,\theta)}(P(r^2+a^2)^2 - Qa^2\sin^2\theta)} J] + Am, \end{aligned} \tag{52}$$

$$0 = B[\sqrt{\frac{\Omega^2 P}{\rho^2}} W_\theta(r,\theta) + \frac{\iota\rho\Omega}{\sin\theta(\sqrt{P(r^2+a^2)^2 - Qa^2\sin^2\theta}} J], \tag{53}$$

$$\begin{aligned}0 &= A[\frac{-E}{\sqrt{F(r,\theta)}} - \sqrt{\frac{\Omega^2 Q}{\rho^2}} W'(r,\theta) \\ &+ \frac{a(P(r^2+a^2)-Q)}{\sqrt{F(r,\theta)}(P(r^2+a^2)^2 - Qa^2\sin^2\theta)} J] + Bm, \end{aligned} \tag{54}$$

$$0 = A[\sqrt{\frac{\Omega^2 P}{\rho^2}} W_\theta(r,\theta) + \frac{\iota\rho\Omega}{\sin\theta\sqrt{P(r^2+a^2)^2 - Qa^2\sin^2\theta}} J]. \tag{55}$$

Solving these equations as explained in Ref. [7] we obtain

$$W_+(r) = \int \frac{(E - \Omega_h J)(r_+^2 + a^2)}{2(r - r_+)(r_+ - M)(1 - \alpha^2 r_+^2)}, \tag{56}$$

which on integration around the pole $r = r_+$ gives

$$W(r) = \pi\iota \frac{(E - \Omega_h J)(r_+^2 + a^2)}{2(r_+ - M)(1 - \alpha^2 r_+^2)}, \tag{57}$$

$$ImW = \pi \frac{(E - \Omega_h J)(r_+^2 + a^2)}{2(r_+ - M)(1 - \alpha^2 r_+^2)}. \tag{58}$$

So, the emission and absorption probabilities of fermions become

$$P_{emission} = \exp[-2ImI] = \exp[-2(ImW_+ + Im\Theta)], \tag{59}$$

$$P_{absorption} = \exp[-2ImI] = \exp[-2(ImW_- + Im\Theta)]. \tag{60}$$

Since $ImW_+ = -ImW_-$, we see that the total probability that the particle tunnels from inside the event horizon to outside is

$$\Gamma \sim \frac{P_{emission}}{P_{absorption}} = \exp[-4ImW_+], \tag{61}$$

or

$$\Gamma = \exp[\frac{-2\pi(E - \Omega_h J)(r_+^2 + a^2)}{(r_+ - M)(1 - \alpha^2 r_+^2)}]. \tag{62}$$

Comparing this with the Boltzmann factor $\Gamma = \exp[-\beta E]$ where $\beta = 1/T_h$ we find that the Hawking temperature [24, 25] is given by

$$T_h = \frac{(r_+ - M)\left(1 - \alpha^2 r_+^2\right)}{2\pi\left(r_+^2 + a^2\right)}, \tag{63}$$

where r_+ is the event horizon. We note that formulae (62) and (63) reduce to the tunneling probability and temperature of the Kerr black hole [26] if acceleration vanishes. Further, we can recover relations for the Schwarzschild black hole if angular momentum also vanishes.

Proceeding as before we work out the tunneling probability at the acceleration horizon, $r_\alpha = 1/\alpha$, and find that the Hawking temperature at this horizon comes out to be

$$T_h = \frac{\alpha\left(\alpha^2 a^2 - 2M\alpha + 1\right)}{2\pi(1 + \alpha^2 a^2)}. \tag{64}$$

4. Hawking radiation of charged fermions

Now we will solve the Dirac equation for accelerating and rotating black holes with electric and magnetic charges. The covariant Dirac equation with charge q is given by

$$i\gamma^\mu\left(D_\mu - \frac{iq}{\hbar}A_\mu\right)\Psi + \frac{m}{\hbar}\Psi = 0, \tag{65}$$

where m is the mass of the fermion particles.

Taking an *ansatz* similar to the one in the previous section for uncharged fermions and following the same steps we get the following set of four equations to be solved.

$$\begin{aligned} 0 &= -B\left[\frac{1}{\sqrt{F(r,\theta)}}\partial_t I_\uparrow + \sqrt{g(r,\theta)}\partial_r I_\uparrow + \frac{H(r,\theta)}{K(r,\theta)\sqrt{F(r,\theta)}}\partial_\phi I_\uparrow \right. \\ &\quad \left. - \frac{1}{\sqrt{F(r,\theta)}}qA_t - \frac{H(r,\theta)}{K(r,\theta)\sqrt{F(r,\theta)}}qA_\phi\right] + mA, \end{aligned} \tag{66}$$

$$0 = -B\left[\frac{1}{\sqrt{\Sigma(r,\theta)}}\partial_\theta I_\uparrow + \frac{i}{\sqrt{K(r,\theta)}}\partial_\phi I_\uparrow - \frac{i}{\sqrt{K(r,\theta)}}qA_\phi\right], \tag{67}$$

$$\begin{aligned} 0 = A&\left[\frac{1}{\sqrt{F(r,\theta)}}\partial_t I_\uparrow - \sqrt{g(r,\theta)}\partial_r I_\uparrow + \frac{H(r,\theta)}{K(r,\theta)\sqrt{F(r,\theta)}}\partial_\phi I_\uparrow \right. \\ &\left. - \frac{1}{\sqrt{F(r,\theta)}}qA_t - \frac{H(r,\theta)}{K(r,\theta)\sqrt{F(r,\theta)}}qA_\phi\right] + mB, \end{aligned} \tag{68}$$

$$0 = -A\left[\frac{1}{\sqrt{\Sigma(r,\theta)}}\partial_\theta I_\uparrow + \frac{i}{\sqrt{K(r,\theta)}}\partial_\phi I_\uparrow - \frac{i}{\sqrt{K(r,\theta)}}qA_\phi\right]. \tag{69}$$

Solving these as before we get the probabilities of crossing the horizon in each direction as [8]

$$P_{emission} \propto \exp\left[-2ImI\right] = \exp\left[-2\left(ImR_+ + Im\Theta\right)\right], \tag{70}$$

$$P_{absorption} \propto \exp\left[-2ImI\right] = \exp\left[-2\left(ImR_- + Im\Theta\right)\right]. \tag{71}$$

Thus the probability of a particle tunneling from inside to outside the horizon is

$$\Gamma \propto \frac{P_{emission}}{P_{absorption}} = \frac{\exp\left[-2\left(ImR_+ + Im\Theta\right)\right]}{\exp\left[-2\left(ImR_- + Im\Theta\right)\right]},$$

or

$$\Gamma = \exp\left[-2\left(ImR_+ - ImR_-\right)\right],$$

which in terms of the event horizon becomes

$$\Gamma = \exp\left[\frac{-2\pi\left(r_+^2 + a^2\right)}{\left(r_+ - M\right)\left(1 - \alpha^2 r_+^2\right)}\left(E - \Omega_h J - \frac{qer_+}{\left(r_+^2 + a^2\right)}\right)\right]. \tag{72}$$

Comparing this with the Boltzmann factor as before we find the Hawking temperature as

$$T_h = \frac{\left(r_+ - M\right)\left(1 - \alpha^2 r_+^2\right)}{2\pi\left(r_+^2 + a^2\right)}. \tag{73}$$

Here again if acceleration vanishes we recover the tunneling probability and temperature for the Kerr-Newman black hole [26]. Taking rotation equal to zero yields the expression for the Reissner-Nordström black hole.

5. Hawking radiation of scalar particles

Black holes emit scalar particles besides other particles [27]. Now we study contribution of scalar fields towards Hawking radiation for the class of black holes at hand. For this purpose the Klein-Gordon equation

$$g^{\mu\nu}\partial_\mu\partial_\nu\Phi - \frac{m^2}{\hbar^2}\Phi = 0, \tag{74}$$

for the scalar field Φ and mass of the scalar particle m will be solved [9]. In order to apply WKB approximation we assume the following solution

$$\Phi\left(t, r, \theta, \phi\right) = \exp\left[\frac{\iota}{\hbar}I\left(t, r, \theta, \phi\right) + I_1\left(t, r, \theta, \phi\right) + O\left(\hbar\right)\right]. \tag{75}$$

Using symmetries of the background spacetime we take the following separation of variables in the action

$$I = -Et + W\left(r\right) + J\phi, \tag{76}$$

where E and J are the energy and angular momentum of the emitted particle. Using this action the Klein-Gordon equation at the horizon, $r = r_+$, takes the form

$$0 = -\frac{1}{F(r_+, \theta)}\left(E - \frac{H(r_+, \theta)}{K(r_+, \theta)}J\right)^2 + \frac{J^2}{K(r_+, \theta)} + g(r_+, \theta)(W'(r))^2 + m^2. \tag{77}$$

Using Eqs. (6) to (10) and (15) in this, and solving by integration gives [9]

$$W_+(r) = \frac{\iota\pi(r_+^2 + a^2)\left(E - \Omega_h J\right)}{2(r_+ - M)(1 - \alpha^2 r_+^2)}, \tag{78}$$

$$ImW_+(r) = \frac{\pi(r_+^2 + a^2)\left(E - \Omega_h J\right)}{2(r_+ - M)(1 - \alpha^2 r_+^2)}. \tag{79}$$

The resulting tunneling probability becomes

$$\Gamma = exp[-2\pi\frac{(r_+^2 + a^2)\left(E - \Omega_h J\right)}{(r_+ - M)(1 - \alpha^2 r_+^2)}]. \tag{80}$$

As before we compare this with the Boltzmann factor to write the Hawking temperature which is the same as in Eq. (73). We again note that the results for *simpler* black holes can be recovered from this. If we put acceleration equal to zero the temperature for the Kerr-Newman black hole is obtained [26]. If α, a, e and g vanish we get Hawking temperature for the Schwarzschild black hole [28] as $T_h = 1/8\pi M$.

For charged scalar particles the outer horizon for accelerating and rotating charged black holes is $r_+ = M + \sqrt{M^2 - a^2 - e^2 - g^2}$. The Klein-Gordon equation for the scalar field Φ with charge q is given by

$$g^{\mu\nu}\left(\partial_\mu - \frac{iq}{\hbar}A_\mu\right)\left(\partial_\nu - \frac{iq}{\hbar}A_\nu\right)\Phi - \frac{m^2}{\hbar^2}\Phi = 0, \tag{81}$$

where A_μ is the vector potential. Using an *ansatz* similar to the one in the uncharged case, the above equation becomes

$$g^{\mu\nu}\left(\partial_\mu I - qA_\mu\right)\left(\partial_\nu I - qA_\nu\right) + m^2 = 0. \tag{82}$$

Following the same procedure as before we obtain the tunneling probability as

$$\Gamma = \exp\left[-\frac{2\pi\left(r_+^2 + a^2\right)}{\left(r_+ - M\right)\left(1 - \alpha^2 r_+^2\right)}\left(E - \Omega_h J - \frac{qer_+}{\left(r_+^2 + a^2\right)}\right)\right], \tag{83}$$

which is the same as in the case of Dirac particles [8]. As before, the Hawking temperature is given by

$$T_h = \frac{\left(r_+ - M\right)\left(1 - \alpha^2 r_+^2\right)}{2\pi\left(r_+^2 + a^2\right)}. \tag{84}$$

The value of r_+ in this case is given by Eq. (12).

6. Conclusion

The Plebański-Demiański family of solutions of Einstein's field equations contains many well-known black hole solutions. These, among others, include the Schwarzschild, Reissner-Nordström, Kerr, Kerr-Newman and Kerr-Newman-NUT solutions. Thus the general results obtained for this family reduce to these particular cases. The family also contains solutions for accelerating and rotating black holes. Here we have reviewed the thermodynamic properties of these solutions. We see that in the case of the Kerr-Newman-NUT black hole the behaviour of temperature is similar to that of the Kerr-Newman black hole if the NUT parameter has less magnitude than that of the sum of the electric and magnetic charges and the rotation parameter. For accelerating and rotating black holes there is a limit on the magnitude of acceleration as well.

Various techniques have been used to study Hawking radiations from black holes. This has been done, for example, by using the Newman-Penrose formalism, the null-geodesic and the so-called Hamilton-Jacobi methods. Here we have used the method of quantum tunneling, which is a variant of the Hamilton-Jacobi method, and complex path integration technique. Radiations from black holes are considered as a phenomenon of quantum tunneling across their event horizons. Using these semi-classical methods we have studied the radiation of charged and uncharged Dirac particles and scalar particles from the Plebański-Demiański class of black holes. The tunneling probabilities for the incoming and outgoing particles have been calculated by solving the Dirac and Klein-Gordon equations in the framework of WKB approximation.

Our procedure works out Hawking temperature for these black configurations as well. Apart from giving results for the case of accelerating and rotating black holes, our formulae reduce to those for the well-known black holes also. We get formulae for the Kerr-Newman black hole when

Journal of Physics: Conference Series **354** (2012) 012016 doi:10.1088/1742-6596/354/1/012016

the NUT parameter vanishes, and the Kerr black hole, when the electric and magnetic charges vanish too. When angular momentum is also zero the thermodynamical quantities reduce to those of the simplest case of the Schwarzschild black hole.

References

[1] Plebański J F and Demiański M 1976 *Ann. Phys.* **98** 98
[2] Podolský J and Kadlecová H 2009 *Class. Quantum Grav.* **26** 105007
[3] Griffiths J B and Podolský J 2005 *Class. Quantum Grav.* **22** 3467
[4] Dias O J C and Lemos J P S 2003 *Phys. Rev.* **D 67** 064001
[5] Dias O J C and Lemos J P S 2003 *Phys. Rev.* **D 67** 084018
[6] Kerner R and Mann R B 2008 *Class. Quant. Grav.* **25** 095014
[7] Gillani U A and Saifullah K 2011 *Phys. Lett.* **B 699** 15
[8] Rehman M and Saifullah K 2011 *JCAP* **03** 001
[9] Gillani U A, Rehman M and Saifullah K 2011 *JCAP* **06** 016
[10] Podolský J and Griffiths j B 2006 *Phys. Rev.* **D 73** 044018
[11] Wald R 1984 *General Relativity*, University of Chicago Press
[12] Ghezelbash A M, Mann R B and Sorkin R D 2007 *Nucl. Phys. B* **775** 95
[13] Bilal M and Saifullah K *Thermodynamics of accelerating and rotating black holes*, arXiv:1010.5575
[14] Hawking S W 1975 *Commun. Math. Phys.* **43** 199
[15] Hawking S W 1974, *Nature* **248** 30
[16] Parikh M K and Wilczek F 2000 *Phys. Rev. Lett.* **85** 5042
[17] Parikh M K 2004 *Gen. Rel. Grav.* **36** 2419
[18] Akhmedova V, Pilling T, de Gill A and Singleton D 2008 *Phys. Lett.* **B 666** 269
[19] Akhmedov E T, Akhmedova V and Singleton D 2006 *Phys. Lett.* **B 642** 124
[20] Umetsu K 2010 *Int. J. Mod. Phys.* **A 25** 4123
[21] Umetsu K 2010 *Phys. Lett.* **B 692** 61
[22] Iso S, Umetsu H and Wilczek F 2006 *Phys. Rev.* **D 74** 044017
[23] Vagenas E C and Das S 2006 *JHEP* **10** 025
[24] Srinivasan K and Padmanabhan T 1999 *Phys. Rev.* **D 60** 24007
[25] Shankaranarayanan S, Padmanabhan T and Srinivasan K 2002 *Class. Quant. Grav.* **19** 2671
[26] Kerner R and Mann R B 2006 *Phys. Rev.* **D 73** 104010
[27] Yale A 2011 *Phys. Lett.* **B 697** 398
[28] Carlip S 2009 in *Physics of Black Holes: A Guided Tour* E. Papantonopoulos (Ed.), Springer-Verlag, Berlin Heidelberg

Journal of Physics: Conference Series **354** (2012) 012017 doi:10.1088/1742-6596/354/1/012017

Accretion of Phantom energy onto a Rotating BTZ Black Hole

Salahuddin
Center for Advanced Mathematics and Physics, National University of Sciences and Technology, H-12, Islamabad, Pakistan
E-mail: salahswati@yahoo.com

Azad A. Siddiqui
Center for Advanced Mathematics and Physics, National University of Sciences and Technology, H-12, Islamabad, Pakistan
E-mail: azad@camp.nust.edu.pk

M. Akbar
Center for Advanced Mathematics and Physics, National University of Sciences and Technology, H-12, Islamabad, Pakistan
E-mail: ak64bar@yahoo.com

Abstract. In this paper, we study the accretion of phantom energy onto a (2+1)-dimension rotating Banados-Teitelboim-Zanelli (BTZ) black hole. An interesting finding is that the rate of change of mass of the rotating BTZ black hole depends not only on the energy density and pressure of the phantom energy but also on the mass of the rotating BTZ black hole.

1. Introduction

In Astrophysics accretion is defined as [1] "a process by which matter is collected around a central object". In binary system accretion, one star is tidally deformed and matter flows out from it to the compact companion. When one deals with an isolated object, it may accrete from the interstellar medium at a very low rate. In many of the galactic centers there is evidence of supermassive black holes. There are no companions, but matter is accreted from winds of surrounding stars. In these cases, stars may also be tidally disrupted if they come very close to the black hole and the matter would be accreted from the disrupted stars to the central black hole.

The recent observational evidence obtained from Wilkinson Microwave Anisotropy Probe (WMAP) strongly suggests that the current expansion of the Universe is accelerating [2]. This accelerated expansion of the Universe is explained by the dominance of dark energy with a negative pressure in Einstein's theory of gravity [3]. A peculiar property of cosmological models with dark energy is the possibility of a Big Rip [4]: an infinite increase in the scale factor of the Universe in a finite time. The Big Rip scenario is realized in the case of phantom energy (for which $(\rho + p) < 0$). In the Big Rip scenario the phantom energy tends to infinity and all

Journal of Physics: Conference Series **354** (2012) 012017 doi:10.1088/1742-6596/354/1/012017

the bound objects are torn apart upto subatomic scales. It should be noted, however, that the condition $(\rho + p) < 0$ alone is not enough for the Big Rip scenario to be realized [5].

The accretion of dark energy onto a black hole has been studied by many authors [6] after the seminal work of Babichev et al [7] who found that

$$dM = 4\pi A M^2 (\rho + p)\, dt, \tag{1}$$

which shows that the mass of the black hole increases as it accretes the gas of particles when $(\rho + p) > 0$, but decreases as it accretes the phantom energy. In particular this implies that the black hole mass in the Universe filled with phantom energy must decrease [7].

The accretion of phantom energy onto a (2+1)-dimensional stationary BTZ black hole was studied by M. Jamil and M. Akbar [8]. In their work they showed that the change in BTZ black hole mass is

$$dM = 2\pi A_1 (\rho + p) dt. \tag{2}$$

Note that for phantom energy $(\rho + p) < 0$, which leads to the decrease in the mass of the BTZ black hole. In this paper we extend this work to rotating BTZ black hole.

2. Model of Accretion for Rotating BTZ Black Hole

The axial symmetric rotating solution to the (2+1)-dimensional Einstein field equations is represented by the rotating BTZ black hole metric [9]

$$ds^2 = -N(r)dt^2 + \frac{1}{N(r)}dr^2 + r^2\left(d\varphi - \frac{J}{2r^2}dt\right)^2, \tag{3}$$

where $N(r) = -M + \frac{r^2}{l^2} + \frac{J^2}{4r^2}$ is the lapse function and M is the dimensionless mass, while J is the angular momentum of the rotating BTZ black hole. In order to discuss the accretion dynamics of phantom energy onto the rotating BTZ black hole we follow the formulism from the work of Babichive et al [7]. Considering the phantom energy to be a perfect fluid for which the energy momentum tensor is

$$T^{ab} = (\rho + p)u^a u^b + p g^{ab}, \tag{4}$$

where ρ and p are the energy density and pressure of the phantom energy respectively, while $u^a = (u^0, u^1, 0)$ is the three vector velocity of the fluid with pressure $p = p(r)$ and energy density $\rho = \rho(r)$. The normalization condition of the fluid velocity implies

$$g_{ab}u^a u^b = -1. \tag{5}$$

Here we consider $u^1 = u$, which is the radial component of velocity of the fluid flow. Solving equation (5) for u^0, gives

$$u^0 = \sqrt{\frac{N(r) + u^2}{N(r)\left(N(r) - \frac{J^2}{4r^2}\right)}}. \tag{6}$$

There are two important equations of conservation on which the accretion process mainly depends. One is the energy flux equation, $T^{0a}{}_{;a} = 0$, and the other is the conservation of mass flux equation, $J^a{}_{;a} = 0$, where J is the current density of the fluid. Since the black hole is stationary, the energy flux conservation equation leads to

$$(\rho+p)ur^3\sqrt{\frac{N(r)\,(N(r) + u^2)}{N(r) - \frac{J^2}{4r^2}}}\left(\frac{\sqrt{M^2 - \frac{J^2}{l^2}} + (M - \frac{2r^2}{l^2})}{\sqrt{M^2 - \frac{J^2}{l^2}} - (M - \frac{2r^2}{l^2})}\right)^{\frac{M}{2\sqrt{M^2 - \frac{J^2}{l^2}}}} \frac{1}{\sqrt{r^2 l^2 (M - \frac{r^2}{l^2}) - \frac{J^2 l^2}{4}}} = C, \tag{7}$$

Journal of Physics: Conference Series **354** (2012) 012017 doi:10.1088/1742-6596/354/1/012017

where C is an arbitrary positive constant. In order to find the second integral of motion we use the energy momentum conservation along the three vector velocity (the energy flux equation)

$$u_a T^{ab}{}_{;b} = 0. \tag{8}$$

Considering the phantom energy to be a perfect fluid, the conservation law becomes [10]

$$\rho_{,b} u^b + (\rho + p) u^b{}_{;b} = 0. \tag{9}$$

The solution of equation (9) is

$$ur\ exp\left(\int_{\rho_\infty}^{\rho_h} \frac{d\rho}{\rho + p}\right) = -A, \tag{10}$$

where A is an arbitrary positive constant. Also ρ_h and ρ_∞ are the energy density of the phantom energy at BTZ black hole horizon and at infinity respectively. Substituting equation (10) in equation (7), we have

$$(\rho+p)r^2\ exp\left[-\int_{\rho_\infty}^{\rho_h} \frac{d\rho}{\rho+p}\right]\sqrt{\frac{N(r)(N(r)+u^2)}{N(r)-\frac{J^2}{4r^2}}}\left(\frac{\sqrt{M^2-\frac{J^2}{l^2}}+(M-\frac{2r^2}{l^2})}{\sqrt{M^2-\frac{J^2}{l^2}}-(M-\frac{2r^2}{l^2})}\right)^{\frac{M}{2\sqrt{M^2-\frac{J^2}{l^2}}}} \times\frac{1}{\sqrt{r^2l^2(M-\frac{r^2}{l^2})-\frac{J^2l^2}{4}}} = C_1, \tag{11}$$

where

$$C_1 = \frac{-C}{A} = \rho_\infty + p_\infty < 0, \tag{12}$$

ρ_∞ is defined above and p_∞ is the pressure at infinity. The rate of change in the mass of rotating BTZ black hole is [8]

$$\frac{dM}{dt} = 2\pi r T_0^1. \tag{13}$$

The value of T_0^1 from the stress energy tensor is obtained as

$$T_0^1 = (\rho+p)u\sqrt{N(r)+u^2}\sqrt{\frac{N(r)-\frac{J^2}{4r^2}}{N(r)}}. \tag{14}$$

Using equations (10), (11), (12) and (14) in equation (13), for general ρ and p, we finally have

$$dM = 2\pi\frac{l}{r}A(\rho+p)\left(\frac{M-\frac{r^2}{l^2}}{(-N(r))^\alpha}\right)\left(\frac{\left(\sqrt{M^2-\frac{J^2}{l^2}}-(M-\frac{2r^2}{l^2})\right)^2}{\frac{4r^2}{l^2}}\right)^{\frac{M}{2\sqrt{M^2-\frac{J^2}{l^2}}}} dt, \tag{15}$$

where

$$\alpha = \frac{\sqrt{M^2-\frac{J^2}{l^2}}+M}{2\sqrt{M^2-\frac{J^2}{l^2}}}. \tag{16}$$

Note that from equation (15) mass of the rotating BTZ black hole decreases for α to be a positive even integer, $M > \frac{r^2}{l^2}$ and $M^2 > \frac{J^2}{l^2}$.

Critical Accretion

At the critical point of accretion the phantom energy shows a variety of changes in its behavior close to the compact object. In order to observe these changes we determine solutions that pass through the critical points. Such solutions correspond to the material falling into the black hole with monotonically increasing speed. To evaluate the critical point of accretion we consider the equation of mass flux or the continuity equation

$$J^a{}_{;a} = 0, \tag{17}$$

which gives

$$\rho u r = C_2, \tag{18}$$

where C_2 is the constant of integration. From equations (18) and (7), we have

$$\left(\frac{\rho+p}{\rho}\right)^2 r^4 \frac{N(r)\left(N(r)+u^2\right)}{N(r)-\frac{J^2}{4r^2}} \left(\frac{\sqrt{M^2-\frac{J^2}{l^2}}+(M-\frac{2r^2}{l^2})}{\sqrt{M^2-\frac{J^2}{l^2}}-(M-\frac{2r^2}{l^2})}\right)^{\frac{M}{\sqrt{M^2-\frac{J^2}{l^2}}}} \frac{1}{r^2l^2(M-\frac{r^2}{l^2})-\frac{J^2l^2}{4}} = C_3, \tag{19}$$

where $C_3 = (\frac{C}{C_2})^2$. Taking the differential of equations (18) and (19) and eliminating $\frac{d\rho}{\rho}$, we get

$$\left[-v^2+\frac{u^2}{N(r)+u^2}\right]\frac{du}{u}+\left[-v^2+\frac{\frac{r^2}{l^2}-\frac{J^2}{4r^2}}{N(r)+u^2}-\frac{M}{-M+\frac{r^2}{l^2}}+\frac{M}{f(r)}\right]\frac{dr}{r}=0, \tag{20}$$

where

$$v^2 = \frac{d\ln(\rho+p)}{d\rho} - 1. \tag{21}$$

It is clear from equation (20) that if one or the other bracket factor vanishes, we get a turn-around point corresponding to double valued solution in either r or u. In order to get the critical points of accretion we take both the bracket factors in equation (20) equal to zero which gives the critical points of accretion as

$$v_c^2 = \frac{\frac{r_c^2}{l^2}-\frac{J^2}{4r_c^2}}{N(r_c)+u^2} - \frac{M}{-M+\frac{r_c^2}{l^2}} + \frac{M}{N(r_c)}, \tag{22}$$

and

$$v_c^2 = \frac{u_c^2}{N(r_c)+u_c^2}, \tag{23}$$

where the subscript c refers to the critical quantities. Comparing equations (22) and (23), we have

$$u_c^2 = \frac{r_c^2}{l^2}\left(\frac{(-M+\frac{r_c^2}{l^2})^2-(\frac{J^2}{4r_c^2})^2}{(-M+\frac{r_c^2}{l^2})^2+\frac{J^2}{4l^2}}\right). \tag{24}$$

In equation (24) the quantity u_c represents the critical speed of flow at the critical point. Furthermore by taking $J = 0$ in equations (23) and (24), one gets the corresponding results of the non-rotating BTZ black hole [8]. For physically acceptable solution, we require $v_c^2 > 0$, hence we get the following restriction on speed and location of critical points

$$u_c^2 > M - \frac{r_c^2}{l^2} - \frac{J^2}{4r_c^2}, \tag{25}$$

and

$$r_c^2 > \frac{l^2M}{2}. \tag{26}$$

Conditions (25) and (26) reduce to the conditions for non-rotating BTZ black hole [8] by taking $J = 0$.

Conclusion

In this paper accretion of phantom energy onto a rotating BTZ spacetime is discussed. We follow the same procedure as followed for the non-rotating BTZ black hole [8]. Our analysis shows that the evolution of the rotating BTZ black hole mass depends not only on the energy density and pressure of the phantom energy but also on the mass of the rotating BTZ black hole. The mass of the rotating BTZ black hole decreases if $\alpha = \frac{\sqrt{M^2-\frac{J^2}{l^2}}+M}{2\sqrt{M^2-\frac{J^2}{l^2}}}$ is a positive even integer, $M > \frac{r^2}{l^2}$ and $M^2 > \frac{J^2}{l^2}$. The physical cause for the decrease in the black hole mass is as follows: the phantom energy falls to the black hole, but the energy flux associated with this fall is directed away from the back hole. We also discussed the critical accretion and found the critical parameters. All the results presented in [8] can be obtained as a special case of our analysis by taking $J = 0$.

References

[1] S.K. Chakrabarti, Study of accretion processes on black holes: Fifty years of developments, *Frontiers in Astrophysics' Ed.* **77**, (2002), 145.

[2] A.G. Riess et al., Observational evidence from supernovae for an accelerating universe and a cosmological constant, *The Astronomical Journal.* **116**, (1998), 1009; S. Perlmutter et al., Measurements of omega and lambda from 42 high redshift supernovae, *The Astronomical Journal.* **517**, (1999), 565; C.L. Bennett et al., First year Wilkinson Microwave Anisotropy Probe (WMAP) observations: preliminary maps and basic results, *The Astronomical Journal Supplement Series.* **148**, (2003), 1; M. Tegmark et al., Cosmological parameters from SDSS and WMAP, *Pyhsical Review D.* **69**, (2004), 103501.

[3] N. Bahcall, J.P. Ostriker, S. Parlmutter and P.J. Steinhardt, The cosmological constant and dark energy, *Pyhsical Review D.* **284**, (1999), 1481.

[4] R.R. Caldwell, Dark energy and the accelerating universe, *Pyhsical Letter B.* **545**, (2002), 23-29.

[5] B. Mcinnes, The dS/CFT correspondence and the big smash, *Journal of High Energy Physics.* **0208**, (2002), 029.

[6] M. Jamil et al., Accretion of phantom energy and generalized second law of thermodynamics for Einstein-Maxwell-Gauss-Bonnet black hole, *The European Physical Journal C.* **58**, (2008), 325; P.F. Gonzalez-Diaz., Accretion of holographic dark energy in a general spherically symmetric black hole, *Pyhsical Review D.* **70**, (2004), 063530; R. G. Cai and A. Wang, Black hole formation from collapsing dust fluid in a background of dark energy, *Pyhsical Review D.* **73**, (2006), 063005.

[7] E. Babichev et al., Black hole mass decreasing due to phantom energy accretion, *Pyhsical Review Letters.* **93**, (2004), 021102.

[8] M. Jamil, M Akbar, Generalized second law of thermodynamics for a phantom energy accreting BTZ black hole, *General Relativity and Gravitation.* **43**, (2011), 1061-1068.

[9] M. Banados et al., Black hole in three-dimensional spacetime, *Physical Reviwe Letter.* **69**, (1992), 1849-1851.

[10] C.W. Misner, K.S. Thorne, and J.A. Wheeler, Gravitation, *vol. 2*, Freeman, W.H. Freeman and Co., San Francisco, 1973.

Journal of Physics: Conference Series **354** (2012) 012018 doi:10.1088/1742-6596/354/1/012018

Probing the correlations in composite signals

S. Sargsyan

Center for Cosmology and Astrophysics, Alikhanian National Laboratory and Yerevan State University, Yerevan, Armenia

E-mail: seda@yerphi.am

Abstract. The technique of degree of randomness is used to model the correlations in sequences containing various subsignals and noise. Kolmogorov stochasticity parameter enables to quantify the randomness in number sequences and hence appears as an efficient tool to distinguish the signals. Numerical experiments for a broad class of composite signals of regular and random properties enable to obtain the qualitative and quantitative criteria for the behavior of the descriptor depending on the input parameters typical to astrophysical signals.

1. The method

The correlations functions and the power spectra are common and efficient tools for the study of correlations in the signals. Astrophysical signals typically are superposition of various subsignals, regular and random, by features comparable to each other and of weaker ones, i.e. perturbations or the noise. The procedure of analysing of the needed signal or signals, their separation from the noise is a common problem while dealing with observations and measurements.

The Kolmogorov stochasticity parameter technique enables to quantify the randomness of sequences of number theory or dynamical systems [1, 2, 3, 4, 5].

The technique of the degree of randomness has been applied to the Cosmic Microwave Background (CMB) temperature sky maps and to the X-ray flux data of the clusters of galaxies. The former data were those obtained by the Wilkinson Microwave Anisotropy Probe (WMAP) during 7-year observations [6, 7], while the X-ray data were obtained by XMM-Newton satellite providing a particularly accurate and complete sky survey (see [8, 9]). In the case of CMB, the Kolmogorov function enabled to separate signals of different origin, e.g. the Galactic and non-Galactic ones, and to detect point sources in the CMB maps [10] (see Fig.1). Concerning the X-ray clusters, it was shown that their X-ray images do possess correlation in the pixelized flux data peculiar to the gravitational potential of the galaxy clusters [11]. This technique resembles the methods of the dynamical systems applied to nonlinear problems (e.g. [12]).

A crucial step in these studies is the modeling and analysis of generated systems, which enables to reveal the behavior of the stochasticity parameter in the case of a given signal and then to consider the application of this technique for real signals [13]. Below we represent the results of numerical experiments for a broad class of signals.

Kolmogorov stochasticity parameter is introduced for a sequence $\{X_1, X_2, \ldots, X_n\}$ of real random variable X sorted in growing order $X_1 \leq X_2 \leq \ldots \leq X_n$. Then the theoretical distribution function is [1, 2]

$$F(X) = n \cdot (probability\, of\, the\, event\, x \leq X). \qquad (1)$$

the stochasticity parameter is defined as

$$\lambda_n = \sqrt{n} \sup_x |F_n(x) - F(x)| \tag{2}$$

where the empirical distribution function is

$$F_n(X) = (number\, of\, the\, elements\, x_i\, which\, are\, less\, than\, X),$$

and

$$F_n(X) = \begin{cases} 0, & X < x_1 \\ k/n, & x_k \le X < x_{k+1} \\ 1, & x_n \le X. \end{cases} \tag{3}$$

Then for the limit

$$\lim_{n\to\infty} P\{\lambda_n \le \lambda\} = \Phi(\lambda)\ , \tag{4}$$

where

$$\Phi(\lambda) = \sum_{k=-\infty}^{+\infty} (-1)^k\, e^{-2k^2\lambda^2},\ \Phi(0) = 0,\ \lambda > 0\ , \tag{5}$$

exists at uniform convergence and independent on F.

Figure 1. The Kolmogorov's function Φ for the portion of the 7-year temperature CMB map obtained by WMAP.

2. Random-regular sequences

We consider a broad class of sequences, i.e. those composed of random x_n and regular $y_n = \frac{an \pmod{b}}{b}$ (a, b are prime numbers) sub-sequences within $(0, 1)$

$$z_n = \alpha x_n + (1 - \alpha) y_n. \tag{6}$$

The parameter α varies within [0,1] defining random sequences at 1 and regular ones at 0, so that by mutually fixing the pair a, b we get new regular sequences.

For z_n we have

$$F(X) = \begin{cases} 0, & X \leq 0 \\ \frac{X^2}{2\alpha(1-\alpha)}, & 0 < X \leq \alpha \\ \frac{2\alpha X - \alpha^2}{2\alpha(1-\alpha)}, & \alpha < X \leq 1-\alpha \\ 1 - \frac{(1-X)^2}{2\alpha(1-\alpha)}, & 1-\alpha < X \leq 1 \\ 1, & X > 1. \end{cases} \quad (7)$$

Figure 2 shows the results of the numerical experiments for 100 sequences, each sequence containing 10000 elements. Each sequence is divided into 50 subsequences, i.e. m runs through values $1, ..., 50$, and for each of them the parameter $\Phi(\lambda_n)_m$ is calculated and then the empirical distribution function $G(\Phi)_m$ of these numbers is obtained. When the original sequences are random, this distribution have to be uniform according to Kolmogorov's theorem. To test that, χ^2 for the functions $G(\Phi)_m$ and $G_0(\Phi) = \Phi$ have been calculated, i.e. one parameter χ^2 is calculated for each of the 100×101 sequences. For 100 χ^2 values per each value of α, we obtained the mean and error values for χ^2, i.e. for each pair a, b we have a plot of the dependence of χ^2 on α.

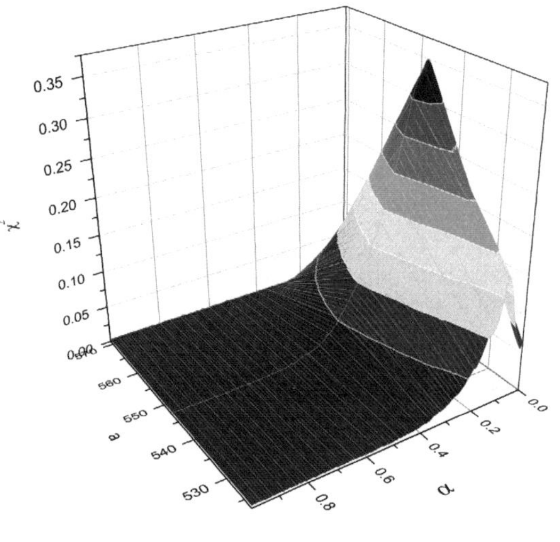

Figure 2. The 3D χ^2 for the Kolmogorov's function for the sequence z_n vs α and the parameter a.

3. Parameters of the regular sequences

At certain values of the parameter a for different values of b the dependences in Fig.2 are monotonic, while for others they do have maxima. To study this effect, we introduce a parameter Δ which is the difference of two values in those plots: maximal value of χ^2 and minimal value in the range $\alpha \in (0, \alpha_{max})$, if α_{max} is the position of the maximal value. Obviously, Δ is zero when the dependence is monotonic and no extrema do exist. Then we calculate Δ for fixed b and for each value of primary $a = 2, ..., b$.

The remarkable feature of the results is the strict mirror symmetry in Fig.3 in the dependence of Δ vs a, although no periodicity is found by Fourier analysis.

The mirror symmetric plots can be hence subjects of particular study, e.g. in two versions: first, of the distribution of Δ and, second, the spacing between non-zero Δ and their distribution.

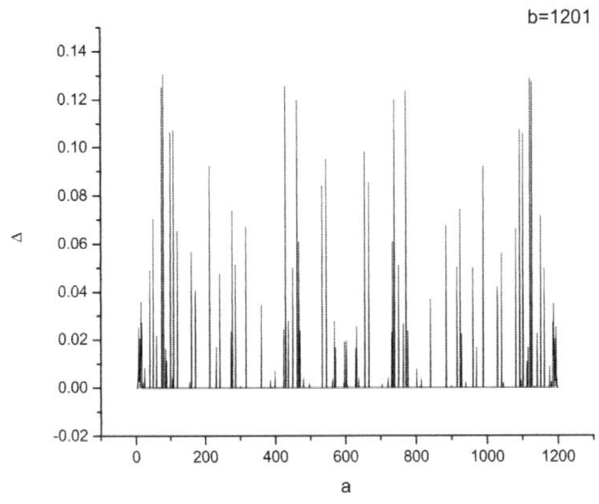

Figure 3. The dependence of Δ vs the parameter a.

The null values of Δ are skipped and also - since due to the mirror symmetry each Δ has its pair - only one of each pair is taken into account.

The results we give in Fig.4 where the number of non-zero Δs from Fig.3 is given in growing order. The number of non-zero Δs appear to be proportional to b.

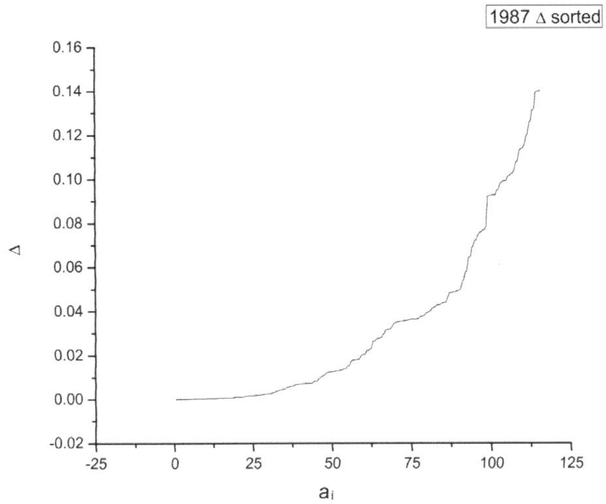

Figure 4. Sorted amplitudes of Δs from Fig.3.

4. Sum of fluctuations: large N limit

The next problem we consider is the properties of the signal being a sum of random and regular fluctuations, each of sequences of 10000 elements and of the same standard deviation.

The regular sequences have been chosen as

$$a_i = \frac{1}{\sqrt{N}} \sum_{k=1}^{N} Compact(x_i^k, -1, 1), \tag{8}$$

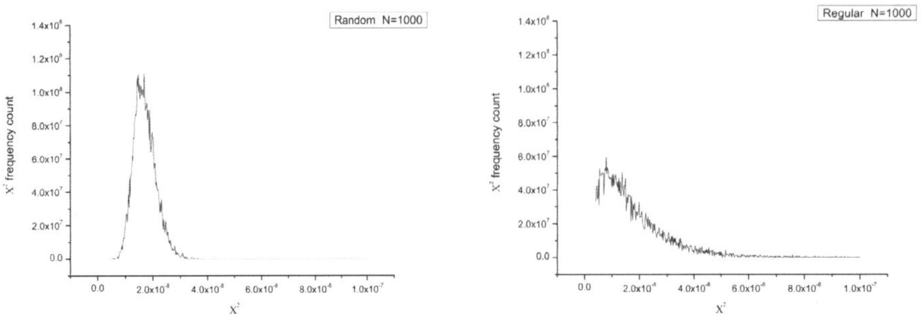

Figure 5. χ^2 frequency for random and regular sequences vs the Gaussian distribution.

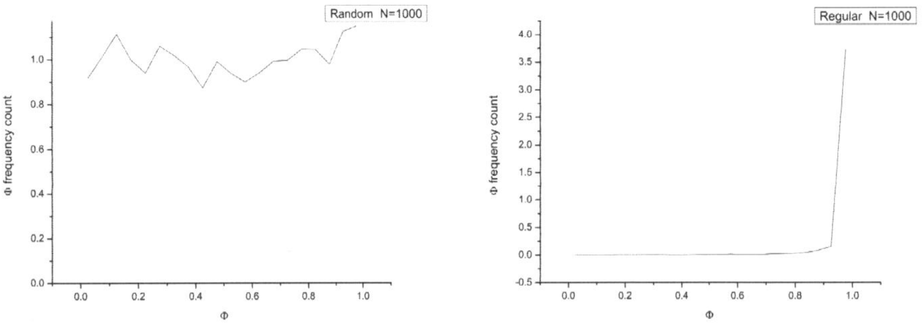

Figure 6. Kolmogorov function Φ for the random and regular sequences in Fig.5.

where

$$x_i^k = i/k,$$

x^k is compactified arithmetical sequence within the interval $(-1, 1)$, with step $1/k$.

The random sequences are given by

$$b_i = \frac{1}{\sqrt{N}} \sum_{k=1}^{N} Random(-1, 1).$$

At large number of sequences each new sequence y_n is taken as the continuation of the former arithmetical progression.

Here

$$Compact(x, p, q) = q + x mod(p - q)$$

indicate multiples of $(p - q)$ from x having the value within the range $(p, q), p < q$.

The results for random and regular sequences, 10000 each, are given in Fig.5 for χ^2, when the number of the fluctuations vary from $N = 1000$ to 100000. The χ^2 shows that for both, random and regular sequences, we deal with a Gaussian limiting distribution, in accordance with the Central Limit theorem which states that for large enough values of N, both sequences a_i and b_i tend to Gaussian sequences with the same σ and μ independent on N. So, although differences are seen in the Gaussians, namely, the standard deviations are larger for the regular case, the χ^2 are similar.

For the Kolmogorov function Φ the situation is rather different. When the Gaussians do appear both for random and regular sequences (as expected), the behaviors of Φ is different and enables to separate them, as shown in Fig.6. Namely, it is close to a homogeneous function for random sequences and $\Phi = 1$ for regular ones. Kolmogorov's function therefore enables to distinguish the superposition of random and regular sequences, even though both are tending to Gaussians.

Finally, we have probed the dependence of the results on the length of the sequences: the dependence on the number of the fluctuations within 1000 – 100000 is rather weak, χ^2 varying around $10^{-8} - 10^{-9}$. This confirms the universality of the obtained behavior of Φ for both random and regular fluctuations.

5. Results

The performed analysis revealed the behavior of the Kolmogorov distribution vs the properties of the generated signals. To describe datasets which contain both regular and stochastic components, we considered sequences scaled by a single parameter α, indicating the ratio of those components.

Quantitative and qualitative criteria have been obtained for the Kolmogorov distribution at numerical experiments for broad class of random and regular sequences depending on α parameter.

a) The existence of the *critical value* for α has been shown, when the monotonic decay of the frequency count of the Kolmogorov distribution is transformed to a function with an extremum.

b) The dependence of scalings and spacings vs that parameter shows mirror properties both in the amplitude and distribution of the frequency counts of the function Φ.

c) The behavior of the randomness of a signal composed of N subsignals at large N limit has been studied, where the Kolmogorov function acts as an informative descriptor. Particularly, the descriptor at large N enables to distinguish the initial set of the fluctuations, even when the superposition both of random and regular subsignals is not informative since tends to a Gaussian in accordance to the Central Limit theorem.

The studied properties are typical, for example, for astrophysical datasets, when the sought signals are superposed with regular and random fluctuations of various origin, and hence the behaviors revealed at the numerical experiments due to the universality of the technique will enable its informative application to real data.

[1] Kolmogorov A.N. 1933 *G.Ist.Ital.Attuari,* **4** 83
[2] Arnold V.I. 2008 *Nonlinearity* **21** T109
[3] Arnold V.I. 2008 *Uspekhi Mat. Nauk* **63** 5
[4] Arnold V.I. 2009 *Trans. Mosc. Math. Soc.* **70** 31
[5] Arnold V.I. 2009 *Funct. Anal. Other Math.* **2** 139
[6] Komatsu E., Dunkley J. *et al.* 2009 *ApJS* **180** 330
[7] Jarosik N., Bennett C.L. *et al.* 2011 *ApJS* **192** 14
[8] Viana P.T.P., da Silva A. *et al.* 2011 *arXiv:1109.1828*
[9] Suhada R., Song J., *et al.* 2011 *arXiv:1111.0141*
[10] Gurzadyan V.G., Allahverdyan A.E. *et al.* 2009 *Astron. & Astrophys.* **497** 343
[11] Gurzadyan V.G., Durret F. *et al.* 2011 *Europhys.Lett.* **95** 69001
[12] Gurzadyan V.G., Pfenniger D., (Eds.) 1994 *Ergodic Concepts in Stellar Dynamics*, Springer-Verlag.
[13] Gurzadyan V.G., Ghahramanyan T., Sargsyan S. 2011 *Europhys.Lett.* **95** 19001

Journal of Physics: Conference Series **354** (2012) 012019 doi:10.1088/1742-6596/354/1/012019

Phantom Energy Accretion by a Class of Black Holes *

M. Sharif †and G. Abbas ‡
Department of Mathematics, University of the Punjab,
Quaid-e-Azam Campus, Lahore-54590, Pakistan.

Abstract

This paper deals with phantom energy accretion by a class of black holes. We give the general formulation for the accretion process by using general static spherically symmetric spacetime. This is then applied to SdS and global charged monopole black holes. We find that the mass of black holes decreases due to phantom accretion. The conditions for critical accretion are also explored which imply a unique critical point in each case.

Keywords: Phantom Energy; Black Holes; Accretion; Critical Accretion.
PACS numbers: 04.70.Bw, 04.70.Dy, 95.35.+d

1 Introduction

Recently, the concept of dark energy (DE) has been recognized by observing that our universe is in accelerating phase. It has been discussed in many cosmological scenarios that the existence of DE with negative pressure leads to a new exotic model of the universe in which it dies as a result of Big Rip [1]-[4]. Also, the anisotropies in cosmic microwave background radiations as

*Talk at *The Third Italian-Pakistani Workshop on Relativistic Astrophysics*, held at Lecce-Italy, June 20-22, 2011.

†msharif.math@pu.edu.pk
‡abbasg91@yahoo.com

observed by WMAP [5]-[7] favor the accelerating behavior of universe. The substance responsible for this behavior is speculated as DE. Due to observational facts, this is a well motivated problem for the theoretical physicists.

Different models like quintessence [8], phantom [9], tachyon field [10], holographic [11] and brane-world [12] help to understand the nature of DE. The simplest form is the cosmological constant for which equation of state (EoS) parameter is $\omega = -1$. The quintessence and phantom are in a hypothetical form of DE for which $\omega > -1$ and $\omega < -1$, respectively [13]-[15]. If $\omega \geqslant -1$, then the DE density is not increasing or decreasing as the universe expands. On the other hand, if we allow $\omega < -1$, then the DE density grows and becomes infinite in a finite time. The expansion of the universe is dominated by the phantom energy which diverges to future singularity (Big Rip).

A massive object surrounded by a matter can capture particles of the matter that passes within certain distance from the massive object. This phenomena is termed as accretion of matter by the massive objects. Bondi [16] originally formulated the problem of matter accretion by the compact objects in Newtonian theory. Michel [17] was the pioneer who studied accretion of gas onto the Schwarzschild black hole (BH) in the relativistic physics. Babichev et al. [18] have shown that BH mass diminishes due to phantom accretion. Jamil et al. [19] have explored the effects of phantom energy accretion onto the charged BH. They pointed out that if the mass of BH becomes smaller (due to accretion of phantom energy) than its charge, then cosmic censorship hypothesis is violated. The same conclusion was deduced by Babichev et al. [20] in a study of the phantom energy accretion onto charged BH with generalized linear EoS.

In the papers [21]-[23], the fate of different BHs in the phantom cosmology has been discussed. In recent papers [24], [25], we have investigated some work about phantom energy accretion. This piece of work is an extension of the previous one for another interesting BH, known as global charged monopole BH by using the procedure of Michel [17]. We also discuss locations of the critical points of accretion. The format of the paper is as follows: In the next section, general formulation is given for spherically symmetric spacetimes. Sections **2** and **3** provide applications to two BHs. Finally, the last section gives the conclusion.

2 General Formalism

A static spherically symmetric BH solution is given by

$$ds^2 = D(r)dt^2 - \frac{1}{D(r)}dr^2 - r^2(d\theta^2 + \sin^2\theta d\phi^2). \tag{1}$$

The horizons of BH can be found by the zero of the lapse function $D(r)$. The energy-momentum tensor for perfect fluid is

$$T_{\mu\nu} = (\rho + p)u_\mu u_\nu - pg_{\mu\nu}, \tag{2}$$

where ρ is the energy density and p is the pressure which violates the null energy condition i.e., $\rho + p < 0$. Also, the only non-zero components of the 4-vector are $u^0 = \frac{dt}{ds}$ and $u^r = \frac{dr}{ds}$ which satisfies the normalization condition, i.e., $u^\mu u_\mu = 1$.

The conservation equation gives

$$r^2 u(\rho + p)\sqrt{D(r) + u^2} = A_0, \tag{3}$$

where A_0 is an integration constant with no dimension and $u^r = u < 0$ for inward flow. Also, by projecting the conservation equation on four-velocity, i.e., $u_\mu \nabla_\nu T^{\mu\nu} = 0$, it follows that

$$r^2 u \exp(n) = -A_1, \tag{4}$$

where $n = \int_{\rho_\infty}^{\rho} \frac{d\rho}{\tilde{\rho}+p(\tilde{\rho})}$ and $A_1 > 0$ is another integration constant (related to the energy flux) implying the dimension of quadratic in length. Here ρ and ρ_∞ are densities of the phantom energy at finite and infinite r. From Eqs.(3) and (4), we get

$$(\rho + p)\left(D(r) + u^2\right)^{\frac{1}{2}} \exp(-n) = A_2, \tag{5}$$

where $A_2 = -\frac{A_0}{A_1} = \rho_\infty + p(\rho_\infty)$. The rate of change of mass due to phantom energy accretion onto $4D$ BH is given by [18]

$$\dot{m} = -4\pi r^2 T^r{}_0. \tag{6}$$

Using Eqs.(3)-(5) in the above equation, it follows that

$$\dot{m} = 4\pi A_1[\rho_\infty + p_\infty]. \tag{7}$$

This implies that the mass of BH increases when realistic matter with $p > 0$ accretes onto the BH. For phantom energy which violates null energy condition $\rho_\infty + p_\infty < 0$, Eq.(7) shows that the accretion of phantom energy onto a BH decreases its mass. The physical reason for the decrease of BH mass lies in the fact that the energy flux associated with the falling phantom energy is always directed away from the BH. The decrease of BH mass is independent of EoS rather than depends only on the violation of null energy condition. It is mentioned here that one can solve Eq.(7) for m by using EoS $p = k\rho$. Since all p and ρ violating null energy condition must satisfy this equation, thus it holds in general.

Now we find such points at which flow speed is equal to the speed of sound. For the discussion of critical points of accretion, we follow the procedure introduced by Michel [17]. The conservation of mass flux, $\nabla_\mu J^\mu = 0$, yields

$$\rho u r^2 = b, \tag{8}$$

where b is an integration constant. Equations (4) and (8) lead to

$$\left(\frac{\rho + p}{\rho}\right)^2 (D(r) + u^2) = b_1, \tag{9}$$

where $b_1 = (\frac{A_0}{b})^2$. Further, Eqs.(8) and (9) yield

$$\frac{dr}{r}\left[2V^2 - \frac{\frac{D'(r)}{2r}}{D(r) + u^2}\right] + \frac{du}{u}\left[V^2 - \frac{u^2}{D(r) + u^2}\right] = 0, \tag{10}$$

where $V^2 = \frac{d\ln(\rho+p)}{d\ln\rho} - 1$ and prime denotes differentiation with respect to r. This equation shows that critical points occur where both the square brackets vanish. The physically acceptable solutions of the above equation are obtained if ${u_c}^2 > 0$ and ${V_c}^2 > 0$.

For the sake of application, we apply the general formulation to SdS as well as global charged monopole BHs.

3 Phantom Energy Accretion by SdS BH

The SdS BH is obtained by taking $D(r) = (1 - \frac{2m}{r} - \frac{r^2}{a^2})$ in Eq.(1), where $a = \sqrt{\frac{3}{\Lambda}}$, m and Λ are constants. This metric has essential singularity at

Journal of Physics: Conference Series **354** (2012) 012019 doi:10.1088/1742-6596/354/1/012019

$r = 0$ covered by the black hole horizons. Such horizons can be found by solving $g_{00} = 1 - \frac{2m}{r} - \frac{r^2}{a^2} \equiv 0$ for r whose positive real roots will give horizons. The corresponding steady state equation of motion (Eq.(10)) becomes

$$\frac{dr}{r}[2V^2 - \frac{\frac{m}{r} + \frac{r^2}{a^2}}{1 - \frac{2m}{r} - \frac{r^2}{a^2} + u^2}] + \frac{du}{u}[V^2 - \frac{u^2}{1 - \frac{2m}{r} - \frac{r^2}{a^2} + u^2}] = 0. \tag{11}$$

For the turn-around points, it follows that

$$u_c{}^2 = \frac{ma^2 + r_c{}^3}{2a^2 r_c}, \quad V_c{}^2 = \frac{ma^2 + r_c^3}{2a^2 r_c - 3ma^2 - r_c{}^3}. \tag{12}$$

The solutions are found when $u_c{}^2 > 0$ and $V_c{}^2 > 0$ implying that

$$2a^2 r_c - 3ma^2 - r_c{}^3 > 0, \quad ma^2 + r_c{}^3 > 0. \tag{13}$$

The cubic equation (13) can be solved (using the technique discussed in [26]) as follows. For $\frac{m}{a} < \frac{4\sqrt{2}}{9\sqrt{3}}$, there are three real roots out of which two are positive and one is negative (neglected). The positive roots are given by

$$r_{c_1} = \frac{2a\sqrt{2}}{\sqrt{3}} \sin\chi, \quad r_{c_2} = \sqrt{2}a(\cos\chi - \frac{1}{\sqrt{3}} \sin\chi), \tag{14}$$

where $\chi = \frac{1}{3}\sin^{-1}(\frac{m}{a}(\frac{3}{2})\frac{5}{2})$. The critical points are found by using Eq.(14) in the 2nd equation of (13) so that

$$\frac{m}{a} = \frac{16}{3}\sqrt{\frac{2}{3}} \sin\chi(\cos^2\chi - 1), \tag{15}$$

and

$$\frac{m}{a} = \frac{2}{27}\sqrt{2}(\sqrt{3}\sin\chi - 3\cos\chi). \tag{16}$$

Since $|\cos^2\chi| \leqslant 1$, so Eq.(15) implies that $\frac{m}{a} \leqslant 0$ - a contradiction, hence r_{c_1} is not a critical point. On the other hand r_{c_2} is a critical solution of flow if $\chi > \frac{\pi}{3}$. When $\frac{m}{a} \geqslant \frac{4\sqrt{2}}{9\sqrt{3}}$, there is no physical solution and hence no critical point.

Journal of Physics: Conference Series **354** (2012) 012019 doi:10.1088/1742-6596/354/1/012019

4 Phantom Energy Accretion by Global Charged Monopole BH

The influence of cosmological phase transitions upon the cosmological defects such as domain wall, strings and monopoles has attracted the attention of the researchers. The monopoles are formed as a result of the gauge symmetry breaking. The simplest model that describes the monopoles is given by the Lagrangian [27]

$$L = \frac{1}{2}\partial_\mu \chi^a \partial^\mu \chi^a - \frac{\lambda}{4}(\chi^a \chi^a - \eta^2)^2, \tag{17}$$

where $\chi^a = \frac{\eta x^a}{8\pi r}$ is isoscalar triplet and $a = 1, 2, 3$. The monopole charge parameter η causes to break the asymptotically flatness of the Schwarzschild solution. A spherically symmetric gravitational collapse of the matter around the monopole forms a BH.

The metric for global charged monopole BH is obtained [28] by replacing $D(r) = (1 - \eta^2 - \frac{2m}{r})$ in Eq.(1), where m is the mass of BH and η is global monopole charge parameter such that $\eta^2 << 1$. This metric has essential singularity at $r = 0$ covered by the BH horizon at $r_h = \frac{2m}{1-\eta^2}$. The equation of motion (10) gives

$$\frac{du}{u}\left[V^2 - \frac{m}{r(1-\eta^2 - \frac{2m}{r} + u^2)}\right] + \frac{dr}{r}\left[2V^2 - \frac{u^2}{(1-\eta^2 - \frac{2m}{r} + u^2)}\right] = 0. \tag{18}$$

This shows that turn-around points (critical points) are located at

$$u_c{}^2 = \frac{m}{2r_c}, \quad V_c^2 = \frac{m}{2r_c(1-\eta^2) - 3m}. \tag{19}$$

The physical solutions exist for $u_c{}^2 \geq 0$ and $V_c^2 \geq 0$ implying that

$$2r_c(1-\eta^2) - 3m \geq 0, \quad m \geq 0. \tag{20}$$

It is worthwhile to mention here that for $\eta \to 0$, the above equations reduce to the results of accretion onto Schwarzschild BH. Further, Eq.(20) implies that

$$r_c = \frac{3m}{2(1-\eta^2)} > 0. \tag{21}$$

The relation between horizon radius and critical point is $r_c = \frac{3}{4}r_h$ which implies that the critical point lies inside the horizons.

Journal of Physics: Conference Series **354** (2012) 012019 doi:10.1088/1742-6596/354/1/012019

5 Outlook

We have formulated the equation of motions for steady state spherically symmetric fluid flow near BH. It has been assumed that infalling fluid does not disturb the generic properties of the BH. Following the procedure introduced by Michel [17], we have derived the relations for the accretion and critical accretion by such a BH for which $g_{11} = (g_{00})^{-1}$. We have applied this formalism to SdS and global charged monopole. For both BHs, we have found that the mass of BHs decreases due to phantom energy accretion on to the BHs. For SdS BH, the critical point can exist only if $\chi > \frac{\pi}{3}$. The critical accretion on the global monopole charged BH implies that $\frac{r_c}{r_h} = \frac{3}{4} < 1$ which indicates that $r_c < r_h$. We conclude that critical point lies inside the horizon of the BH.

Acknowledgment

We would like to thank the Higher Education Commission, Islamabad, Pakistan for its financial support through the *Indigenous Ph.D. 5000 Fellowship Program Batch-IV*.

References

[1] Perlmutter S et al 1997 Astrophys. J. **483** 565

[2] Perlmutter S et al 1998 Nature **391** 51

[3] Perlmutter S et al 1999 Astrophys. J. **517** 565

[4] Riess A G et al 1998 Astron. J. **116** 1009

[5] Bennett C L et al 2003 Astrophys. J. Suppl. **148** 1

[6] Spergel D N et al 2003 Astrophys. J. Suppl. **148** 175

[7] Verde L et al 2002 Mon. Not. R. Astron. Soc. **335** 432

[8] Sahni V and Starobinsky A A 2000 Int. J. Mod. Phys. D **9** 373

[9] Caldwell R R 2002 Phys. Lett. B **23** 545

[10] Sen A J 2002 High Energy Phys. **48** 204

[11] Wang B, Gong Y G and Abdalla E 2005 Phys. Lett. B **624** 141

[12] Li M 2004 Phys. Lett. B **603** 1

[13] Frieman J A et al 1995 Phys. Rev. Lett. **75** 2077

[14] Coble K, Dodelson S and Frieman J A 1997 Phys. Rev. D **55** 1851

[15] Hinshaw G et al 2009 Astrophys. J. Suppl. **180** 225

[16] Bondi H 1952 Mon. Not. Roy. Astron. Soc. **112** 195

[17] Michel F C 1972 Astrophys. Space Sci. **15** 153

[18] Babichev E, Dokuchaev V and Eroshenko Y 2004 Phys. Rev. Lett. **93** 021102

[19] Jamil M, Rashid M and Qadir A 2008 Eur. Phys. J. C **58** 325

[20] E. Babichev, V. Dokuchaev and Y. Eroshenko, J. Exp. Theor. Phys. **112**(2011)784.

[21] Jamil M 2009 Eur. Phys. J. C **62** 609

[22] Jamil M and Akbar M 2011 Gen. Relativ. Gravit. **43** 1061

[23] Jamil M and Qadir A 2011 Gen. Relativ. Gravit. **43** 1069

[24] Sharif M and Abbas G 2011 Chin. Phys. Lett. **28** 090402

[25] Sharif M and Abbas G 2011 Mod. Phys. Lett. A **26** 1731

[26] Sung-Won K 1991 J. Korean Phys. Soc. **24** 118.

[27] Barriola M and Vilenkin A 1989 Phys. Rev. Lett. **63** 341

[28] Zloshchastiev K G 1998 Phys. Rev. D **57**4812

Dynamics of Spherically Symmetric Gravitational Collapse in $f(R)$ Theory

M. Sharif[†] **and H. Rizwana Kausar** [‡]
Department of Mathematics, University of the Punjab,
Quaid-e-Azam Campus, Lahore-54590, Pakistan

E-mail: † msharif.math@pu.edu.pk
‡ rizwa_math@yahoo.com

Abstract. We study dynamics of spherically symmetric gravitational collapse in modified $f(R)$ gravity. The matter filled inside the collapsing body is of two types: a) the usual matter that consists of isotropic non-viscous fluid satisfying the energy conditions; b) the matter for which the effective stress-energy tensor contains higher order curvature derivatives violating the null energy conditions. Thus the higher order curvature terms are interpreted as the gravitational fluid or $f(R)$ dark energy. We discuss the effects of such type of dark energy on the coupling of dynamical and transport equations.

Keywords: $f(R)$ dark energy; Dynamical equations; Transport equation.
PACS: 04.50.Kd

1. Introduction

What is the origin of the perceived dark energy (DE) component which drives the acceleration? This is a well-posed and interesting question which has gained much attraction recently. Several radically different candidates have been proposed to explain it ranging from the introduction of a cosmological constant [1] in General Relativity (GR) to the development of modified theories of gravity [2]. In the latter context, one may assume that at large scales, GR breaks down and a more general function describes the gravitational field. The task of these modifications is to fit the astrophysical data without adding exotic ingredients.

Recently, many generalizations/modifications have been proposed. In particular, there is a considerable interest in generalizing the Ricci scalar curvature R with an arbitrary function of Ricci scalar, $f(R)$, in the Einstein-Hilbert action. This task was initially developed by Buchdahl [3] and further extended by some other people [4]. Such a modification successfully describes the DE epoch and passes the solar system tests [5]-[9]. It is remarkable that even the form of $f(R)$ gravity may be reconstructed from the known universe [10]. Hence, this approach may suggest gravitational alternative for DE.

The action for the modified $f(R)$ theory of gravity is given by

$$S=\frac{1}{2\kappa}\int d^4x\sqrt{-g}f(R). \tag{1}$$

Here the presence of $f(R)$ function may be understood as the introduction of an effective fluid which is not restricted to hold the usual energy conditions [11]. In this way, DE can be thought of as having the geometrical origin rather than some additional scalar fields which are added by hand to the matter part. It is also possible to show that $f(R)$ theory can play a major role at astrophysical scales due to modification of gravitational potential in the low energy limit. Such type of modified potential offers the possibility to fit galaxy rotation and galaxy cluster potential without need of huge amount of dark matter [12].

In this work, we are concerned with spherically symmetric stars to investigate how $f(R)$ dark energy affects the process of gravitational collapse. To carry out this work, we have taken care of the literature [13, 14]. The scheme of the paper is as follows. In next section **2**, we present some basic material useful for this work. Section **3** is devoted to formulate the dynamical equations in $f(R)$ gravity. In section **4**, the transport equation and the coupling with the dynamical equations is given. The last section **5** provides a summary and discussion of the results.

2. $f(R)$ Field Equations

We consider the spherically symmetric line element as follows

$$ds^2 = A^2(t,r)dt^2 - B^2(t,r)dr^2 - C^2(t,r)(d\theta^2 + \sin^2\theta d\phi^2). \tag{2}$$

An isotropic distribution of matter is assumed inside the collapsing sphere which satisfies the energy conditions and undergoes dissipation in the form of heat flux. It is given by

$$T^m_{\alpha\beta} = (\rho + p)u_\alpha u_\beta - p g_{\alpha\beta} + u_\alpha q_\beta + q_\alpha u_\beta, \quad (\alpha, \beta = 0, 1, 2, 3). \tag{3}$$

Here, we have taken ρ as the energy density, p the pressure, q_α the heat flux and u_α the four-velocity of the fluid. These quantities satisfy the relations

$$u^\alpha u_\alpha = 1, \quad u^\alpha q_\alpha = 0. \tag{4}$$

Using the metric approach, i.e., varying the action in Eq.(1) with respect to $g_{\alpha\beta}$ provides the following field equations

$$F(R)R_{\alpha\beta} - \frac{1}{2}f(R)g_{\alpha\beta} - \nabla_\alpha\nabla_\beta F(R) + g_{\alpha\beta}F(R) = \kappa T_{\alpha\beta},$$
$$F(R) \equiv df(R)/dR. \tag{5}$$

By taking all the higher order corrections to the curvature on the right hand side of the field equations and defining it as a "dark source" term or "effective fluid" [15], this equation can be written as

$$G_{\alpha\beta} = \frac{\kappa}{F}(T^{(m)}_{\alpha\beta} + T^{(D)}_{\alpha\beta}), \tag{6}$$

where

$$T^D_{\alpha\beta} = \frac{1}{\kappa}\left[\frac{f(R) + RF(R)}{2}g_{\alpha\beta} + \nabla_\alpha\nabla_\beta F(R) - g_{\alpha\beta}F(R)\right]. \tag{7}$$

This makes it easy to compare $f(R)$ theory with GR. For the metric (2), the field equations (5) take the following form

$$\left(\frac{2\dot{B}}{B} + \frac{\dot{C}}{C}\right)\frac{\dot{C}}{C} - \left(\frac{A}{B}\right)^2\left[\frac{2C''}{C} + \left(\frac{C'}{C}\right)^2 - \frac{2B'C'}{BC} - \left(\frac{B}{C}\right)^2\right] = \frac{\kappa}{F}\left[\rho A^2\right.$$
$$\left. + \frac{A^2}{\kappa}\left\{\frac{f - RF}{2} + \frac{F''}{B^2} + \left(\frac{2\dot{C}}{C} - \frac{\dot{B}}{B}\right)\frac{\dot{F}}{A^2} + \left(\frac{2C'}{C} - \frac{B'}{B}\right)\frac{F'}{B^2}\right\}\right], \tag{8}$$

$$-2\left(\frac{\dot{C}'}{C} - \frac{\dot{C}A'}{CA} - \frac{\dot{B}C'}{BC}\right) = \frac{\kappa}{F}\left[-qAB + \frac{1}{\kappa}\left(\dot{F}' - \frac{A'}{A}\dot{F} - \frac{\dot{B}}{B}F'\right)\right], \tag{9}$$

$$-\left(\frac{B}{A}\right)^2\left[\frac{2\ddot{C}}{C}-\left(\frac{2\dot{A}}{A}-\frac{\dot{C}}{C}\right)\frac{\dot{C}}{C}\right]+\left(\frac{2A'}{A}+\frac{C'}{C}\right)\frac{C'}{C}-\left(\frac{B}{C}\right)^2=\frac{\kappa}{F}\left[pB^2\right.$$
$$\left.-\frac{B^2}{\kappa}\left\{\frac{f-RF}{2}-\frac{\ddot{F}}{A^2}+\left(\frac{\dot{A}}{A}+\frac{2\dot{C}}{C}\right)\frac{\dot{F}}{A^2}+\left(\frac{A'}{A}+\frac{2C'}{C}\right)\frac{F'}{B^2}\right\}\right], \tag{10}$$
$$-\left(\frac{C}{A}\right)^2\left[\frac{\ddot{B}}{B}-\frac{\ddot{C}}{C}-\frac{\dot{A}}{A}\left(\frac{\dot{B}}{B}+\frac{\dot{C}}{C}\right)+\frac{\dot{B}\dot{C}}{BC}\right]+\left(\frac{C}{B}\right)^2\left[\frac{A''}{A}+\frac{C''}{C}-\frac{A'B'}{AB}\right.$$
$$\left.+\left(\frac{A'}{A}-\frac{B'}{B}\right)\frac{C'}{C}\right]=\frac{\kappa}{F}\left[pC^2-\frac{C^2}{\kappa}\left\{\frac{f-RF}{2}-\frac{\ddot{F}}{A^2}+\frac{F''}{B^2}+\frac{\dot{F}}{A^2}\right.\right.$$
$$\left.\left.\times\left(\frac{\dot{A}}{A}-\frac{\dot{B}}{B}+\frac{\dot{C}}{C}\right)+\left(\frac{A'}{A}-\frac{B'}{B}+\frac{C'}{C}\right)\frac{F'}{B^2}\right\}\right], \tag{11}$$

where dot and prime represent derivatives with respect to t and r respectively. The mass function introduced by Misner and Sharp is defined as follows [16]

$$m(t,r)=\frac{C}{2}(1+g^{\mu\nu}C_{,\mu}C_{,\nu})=\frac{C}{2}\left(1+\frac{\dot{C}^2}{A^2}-\frac{C'^2}{B^2}\right) \tag{12}$$

which provides total energy inside a spherical body of radius "C".

3. Dynamical Equations

The dynamical equations can be obtained from the contracted Bianchi identities [14]. Consider the following two equations

$$\left(T^{(m)\alpha\beta}+T^{(D)\alpha\beta}\right)_{;\beta}u_\alpha=0,\quad \left(T^{(m)\alpha\beta}+T^{(D)\alpha\beta}\right)_{;\beta}\chi_\alpha=0 \tag{13}$$

which yield respectively

$$\frac{\dot{\rho}}{A}+\frac{1}{A}(\rho+p)(\frac{\dot{B}}{B}+\frac{2\dot{C}}{C})+\frac{q}{B}\left(\frac{A'}{A}+\frac{2C'}{C}\right)+\frac{q'}{B}+D_1, \tag{14}$$
$$\frac{\dot{q}}{A}+\frac{2q}{A}\left(\frac{\dot{B}}{B}+\frac{\dot{C}}{C}\right)+\frac{A'}{AB}(\rho+p)+\frac{p'}{B}+D_2, \tag{15}$$

where D_1 and D_2 are given in **Appendix**. The proper time and radial derivatives are given by

$$D_T=\frac{1}{A}\frac{\partial}{\partial t},\quad D_C=\frac{1}{C'}\frac{\partial}{\partial r}. \tag{16}$$

The velocity of the collapsing fluid is defined by the proper time derivative of C, i.e.,

$$U=D_TC=\frac{\dot{C}}{A} \tag{17}$$

which is always negative in the case of collapse. Using this definition, we can write Eq.(12) as follows

$$E\equiv\frac{C'}{B}=\left[1+U^2+\frac{2m}{C}\right]^{1/2}. \tag{18}$$

The variation of total energy of the collapsing sphere of radius C with respect to proper time can be obtained by applying Eq.(16) to (12) along with Eqs.(9), (10) and (18) as follows

$$D_Tm\ =\ \frac{-C^2}{2}\left[\left(\frac{\kappa}{F}p-\frac{1}{4C^2}-D_3\right)U+\left(\frac{\kappa q}{F}+D_4\right)E\right] \tag{19}$$

where D_3 and D_4 are written in **Appendix**. In Eq.(19), terms inside the first round brackets increase the energy density through the rate of work being done by the effective pressure p and the DE in D_3 whose negative effect is balanced with collapsing velocity U. The terms in the second round brackets have negative sign indicating that energy is leaving the system due to heat flux and repulsive effect of curvature terms. Similarly, the variation of energy between adjacent spherical surfaces inside the considered matter distribution can be obtained by using Eqs.(8) and (9) as follows

$$D_C m = \frac{C^2}{2}\left[\left(\frac{\kappa q}{F}+D_4\right)\frac{U}{E}+\frac{\kappa\rho}{F}+D_5\right], \tag{20}$$

where D_5 being the dark source term given in **Appendix**. Equation (20) shows the contribution of energy density, heat flux with the coefficients and D_4, D_5 containing higher order curvature corrections (DE) which affect the adjacent layers of collapsing sphere.

The acceleration $D_T U$ of the collapsing matter inside the spherical body is found by using Eqs.(10), (12), (17) and (18) as follows

$$\begin{aligned} D_T U &= \frac{A'}{AB}E-\frac{\kappa}{2F}\left[p-\frac{1}{\kappa}\left\{\frac{f(R)-RF}{2}-\frac{\ddot{F}}{A^2}\right.\right. \\ &+ \left.\left.\frac{\dot{F}}{A^2}\left(\frac{\dot{A}}{A}+\frac{2\dot{C}}{C}\right)+\frac{F'}{B^2}\left(\frac{A'}{A}+\frac{2C'}{C}\right)\right\}\right]. \end{aligned} \tag{21}$$

This equation implies that

$$\begin{aligned} \frac{A'}{AB}(\rho+p) &= \frac{(\rho+p)}{E}D_T U+\frac{(\rho+p)\kappa}{2FE}\left[p-\frac{1}{\kappa}\left\{\frac{f(R)-RF}{2}-\frac{\ddot{F}}{A^2}\right.\right. \\ &+ \left.\left.\frac{\dot{F}}{A^2}\left(\frac{\dot{A}}{A}+\frac{2\dot{C}}{C}\right)+\frac{F'}{B^2}\left(\frac{A'}{A}+\frac{2C'}{C}\right)\right\}\right]. \end{aligned} \tag{22}$$

Replacing this value with the third term of Eq.(15), we get

$$\begin{aligned} (\rho+p)D_T U &= -(\rho+p)D_6-E^2 D_C p \\ &- E\left[2q\left(\frac{D_T B}{B}-\frac{D_T C}{C}\right)+D_T q\right]+D_7, \end{aligned} \tag{23}$$

where D_6 and D_7 are provided in **Appendix**. Equation (23) represents the role of different forces including DE on the collapsing process. It can be interpreted in the form of Newton's second law of motion, i.e., Force = inertial mass $\times$ acceleration. The term $\rho+p$ on the LHS indicates inertial mass density multiplying with acceleration $D_T U$. The first term on the RHS is the same and can be interpreted as an active gravitational mass by the equivalence principle. We see that how this mass is affected by DE. The terms in the second square brackets appear with negative sign enhancing the collapsing rate which show the gradient of the pressure. Therefore this term would increase the rate of collapse. The terms in the last square brackets play the role of heat flux and a strong contribution of DE in D_7 creating repulsive effect on the surroundings, hence reducing the rate of collapse. This leads to a point that spherical body may not collapse fully in any case due to immense of DE terms in Eq.(23), whatsoever gravity pulls towards the center.

4. Transport Equation

Here the transport equation for heat flux is derived using the Müller-Israel-Stewart theory of dissipative fluids [17]. It helps to study the transfer of mass, heat and momentum during the matter collapse. This is given by

$$\tau h^{\alpha\beta}u^{\gamma}q_{\beta;\gamma}+q^{\alpha}=-\eta h^{\alpha\beta}(T_{,\beta}+a_{\beta}T)-\frac{1}{2}\eta T^{2}\left(\frac{\tau u^{\beta}}{\eta T^{2}}\right)_{;\beta}q^{\alpha}, \tag{24}$$

where $h^{\alpha\beta}=g^{\alpha\beta}-u^{\alpha}u^{\beta}$ is the projection tensor, η denotes the thermal conductivity, T is the temperature, τ stands for relaxation time which is the time taken by a perturbed system to return into an equilibrium state and $a_{\beta}T$ is the Tolman inertial term with $a_{\alpha}=u_{\alpha;\beta}u^{\beta}$ being the acceleration. The independent component of this equation is given by

$$\tau\dot{q}=-qA-\frac{1}{2}\eta qT^{2}\left(\frac{\tau}{\eta T^{2}}\right)^{\dot{}}-\frac{1}{2}\tau q\left(\frac{\dot{B}}{B}+2\frac{\dot{C}}{C}\right)+\frac{\eta A^{2}}{B}\left(\frac{T}{A}\right)'. \tag{25}$$

Substituting the value of $\frac{A'}{A}$ from Eq.(22) and using Eqs.(17) and (18), it follows that

$$\begin{aligned} D_{T}q &= -\frac{\eta T^{2}q}{2\tau}D_{T}\left(\frac{\tau}{\eta T^{2}}\right)-\frac{q}{2}\left(\frac{D_{T}B}{B}+\frac{2D_{T}C}{C}+\frac{1}{\tau}\right)+\frac{\eta E}{\tau}D_{C}T\\ &- \frac{\eta T}{\tau E}D_{T}U-\frac{\eta T}{\tau E}D_{6}. \end{aligned} \tag{26}$$

Now we couple this transport equation with dynamical Eq.(23) to see the effects of heat flux on collapsing process. Using Eq.(26) in (23), we obtain

$$\begin{aligned} (\rho+p)(1-\alpha)D_{T}U &= -(1-\alpha)(\rho+p)D_{6}-E\left[-\frac{\eta T^{2}q}{2\tau}D_{T}\left(\frac{\tau}{\eta T^{2}}\right)-E^{2}D_{C}p\right.\\ &- \left.\frac{q}{2}\left(\frac{D_{T}B}{B}+\frac{2D_{T}C}{C}+\frac{1}{\tau}\right)+\frac{\eta E}{\tau}D_{C}T\right]+D_{7}, \end{aligned} \tag{27}$$

where α is given by

$$\alpha=\frac{\eta T}{\tau(\rho+p)}. \tag{28}$$

We see that inertial mass density on the LHS of dynamical equation (27) is affected by the factor $1-\alpha$ arising from Tolman term. The same factor also appears with the gravitational force term on the RHS which shows the consistency of equivalence principle. However, we note that this factor is independent of the curvature fluid terms. It is mentioned here that the evolution process of collapsing sphere now depends upon the values of α and terms inside the first square brackets in Eq.(27). The terms inside the second square brackets and expressions, namely, D_6 representing gravitational force in Eq.(34) and D_7, the DE contributions in Eq.(35), are the same as in dynamical equation (23).

5. Summary

The aim of this paper is to study the problem of gravitational collapse in $f(R)$ theory of gravity which is strongly motivated by the observational discovery of DE. Here the higher order curvature terms are imposed to violate the null energy conditions and thought to be the origin of DE. We have considered spherically symmetric collapsing stars and discussed their dynamics under the influence of $f(R)$ DE. Effects of outgoing heat flux and anisotropic pressure have also been taken into account. Firstly, we have formulated dynamical equations by using contracted Bianchi identities. Secondly, we have developed transport equation and its coupling with dynamical equation.

The dynamical equations help to investigate the evolution of gravitational collapse with time. These equations yield the variation of total energy inside a collapsing body with respect to time and adjacent surfaces, as found in Eqs.(19) and (20) respectively. Here curvature terms appear to affect the passive gravitational mass and rate of collapse. Thus $f(R)$ DE slows down the rate of the collapse due its repulsive effect.

The transport equation governs the dissipative fluxes and their associated quantities like temperature, relaxation time and thermal conductivity. It helps to construct physically viable models of radiating stars. After coupling the dynamical and transport equations, the above quantities affect the dynamics of collapsing process other than $f(R)$ DE. The appearance of additional factor α has a particular effect on the inertial mass density and gravitational force term.

Acknowledgment

We would like to thank the Higher Education Commission, Islamabad, Pakistan for its financial support through the *Indigenous Ph.D. 5000 Fellowship Program Batch-III.*

Appendix

Here, we explicitly give the expressions for the coefficients D_1, D_2, D_3, D_4, D_5, D_6 and D_7 representing the high order terms in the curvature.

$$\begin{aligned}
D_1 &= \frac{A}{\kappa}\left[\left\{\frac{1}{A^2B^2}\left(\dot{F}'-\frac{A'}{A}\dot{F}-\frac{\dot{B}}{B}F'\right)\right\}_{,1}\right.\\
&+ \left\{\frac{f-RF}{2A^2}+\frac{F''}{A^2B^2}-\frac{\dot{F}}{A^2}\left(\frac{\dot{B}}{B}-\frac{2\dot{C}}{C}\right)-\frac{F'}{B^2}\left(\frac{B'}{B}-\frac{2C'}{C}\right)\right\}_{,0}\\
&+ \frac{\dot{A}}{A^3}\left\{\frac{f-RF}{2A^2}+\frac{F''}{A^2B^2}-\frac{\dot{F}}{A^2}\left(\frac{\dot{B}}{B}-\frac{2\dot{C}}{C}\right)-\frac{F'}{B^2}\left(\frac{B'}{B}-\frac{2C'}{C}\right)\right\}\\
&+ \frac{\dot{B}}{BA^2}\left\{\frac{F''}{B^2}+\frac{\ddot{F}}{A^2}-\frac{\dot{F}}{A^2}\left(\frac{\dot{A}}{A}+\frac{\dot{B}}{B}\right)-\frac{F'}{B^2}\left(\frac{A'}{A}+\frac{B'}{B}\right)\right\}\\
&+ \frac{2\dot{C}}{CA^2}\left\{\frac{\ddot{F}}{A^2}+\frac{\dot{F}}{A^2}\left(\frac{\dot{C}}{C}-\frac{\dot{A}}{A}\right)-\frac{F'}{B^2}\left(\frac{A'}{A}-\frac{C'}{C}\right)\right\}\\
&+ \left.\frac{1}{A^2B^2}\left(\dot{F}'-\frac{A'}{A}\dot{F}-\frac{\dot{B}}{B}F'\right)\left(\frac{2A'}{A}+\frac{B'}{B}+\frac{C'}{C}\right)\right], \qquad (29)
\end{aligned}$$

$$\begin{aligned}
D_2 &= -\frac{B}{\kappa}\left[\left\{\frac{1}{A^2B^2}\left(\dot{F}'-\frac{A'}{A}\dot{F}-\frac{\dot{B}}{B}F'\right)\right\}_{,0}\right.\\
&+ \left\{\frac{f-RF}{2B^2}-\frac{\ddot{F}}{A^2}+\frac{\dot{F}}{A^2}\left(\frac{\dot{A}}{A}+\frac{2\dot{C}}{C}\right)+\frac{F'}{B^2}\left(\frac{A'}{A}+\frac{2C'}{C}\right)\right\}_{,1}\\
&+ \frac{A'}{AB^2}\left\{\frac{\ddot{F}}{A^2}+\frac{F''}{B^2}-\frac{\dot{F}}{A^2}\left(\frac{\dot{A}}{A}+\frac{\dot{B}}{B}\right)-\frac{F'}{B^2}\left(\frac{A'}{A}-\frac{B'}{B}\right)\right\}
\end{aligned}$$

$$
\begin{aligned}
&+ \frac{2B'}{B^3}\left\{\frac{f-RF}{2A^2}+\frac{\ddot{F}}{A^2}+\frac{\ddot{F}}{A^2}-\frac{\dot{F}}{A^2}\left(\frac{\dot{A}}{A}+\frac{\dot{B}}{B}\right)-\frac{F'}{B^2}\left(\frac{A'}{A}+\frac{B'}{B}\right)\right\}\\
&+ \frac{2\dot{C}}{CA^2}\left\{\frac{\ddot{F}}{A^2}+\frac{\dot{F}}{A^2}\left(\frac{\dot{C}}{C}-\frac{\dot{A}}{A}\right)-\frac{F'}{B^2}\left(\frac{A'}{A}-\frac{C'}{C}\right)\right\}\\
&+ \frac{1}{A^2B^2}\left(\dot{F}'-\frac{A'}{A}\dot{F}-\frac{\dot{B}}{B}F'\right)\left(\frac{\dot{A}}{A}+\frac{3\dot{B}}{B}+\frac{2\dot{C}}{C}\right)\Bigg], \qquad (30)
\end{aligned}
$$

$$
D_3 = \frac{f(R)-RF}{2F}-\frac{\ddot{F}}{A^2F}+\frac{\dot{F}}{A^2F}\left(\frac{\dot{A}}{A}+\frac{2\dot{C}}{C}\right)+\frac{F'}{FB^2}\left(\frac{A'}{A}+\frac{2C'}{C}\right), \qquad (31)
$$

$$
D_4 = \frac{1}{AB}\left(\frac{A'}{A}\frac{\dot{F}}{F}-\frac{\dot{F}'}{F}+\frac{\dot{B}}{B}\frac{F'}{F}\right), \qquad (32)
$$

$$
D_5 = \frac{F''}{FB^2}+\frac{f(R)-RF}{2F}-\frac{\dot{F}}{FA^2}\left(\frac{\dot{B}}{B}-\frac{2\dot{C}}{C}\right)-\frac{F'}{B^2F}\left(\frac{B'}{B}-\frac{2C'}{C}\right), \qquad (33)
$$

$$
\begin{aligned}
D_6 &= \left[\frac{\kappa p_r}{2F}-\frac{1}{2F}\left\{\frac{f(R)-RF}{2}\right.\right.\\
&- \left.\left. D_T\left(\frac{\dot{F}}{A}\right)+\frac{2D_TFD_TC}{C}+\frac{E}{B}\left(\frac{A'}{A}+\frac{2C'}{C}\right)D_CF\right\}\right], \qquad (34)
\end{aligned}
$$

$$
\begin{aligned}
D_7 &= \frac{E}{\kappa B}\left[\left(\frac{D_TA}{A}-\frac{D_TB}{B}-\frac{2D_TC}{C}\right)\left(D_TF'-\frac{A'}{A}D_TF-\frac{F'}{B}D_TB\right)\right.\\
&+ \left\{\frac{2}{C}D_TFD_TC-D_T\left(\frac{\dot{F}}{A}\right)+\frac{E}{B}\left(\frac{A'}{A}+\frac{2C'}{C}\right)D_CF\right\}_{,1}\\
&+ \frac{A'}{A}\left\{\frac{F''}{B^2}+D_T\left(\frac{\dot{F}}{A}\right)-\frac{D_TFD_TB}{B}-\frac{F'}{B^2}\left(\frac{A'}{A}+\frac{B'}{B}\right)\right\}\\
&+ \frac{2C'}{C}\left\{\frac{F''}{B^2}-D_TF\left(\frac{D_TB}{B}+\frac{D_TC}{C}\right)-\frac{F'}{B^2}\left(\frac{B'}{B}+\frac{C'}{C}\right)\right\}\\
&+ \left.\frac{1}{A^2}\left(\dot{F}'-\frac{A'}{A}\dot{F}-\frac{\dot{B}}{B}F'\right)_{,0}\right]. \qquad (35)
\end{aligned}
$$

References

[1] Weinberg, S.: Rev. Mod. Phys. **61**(1989)1; Peebles, P.J.E. and Ratra, B.: Rev. Mod. Phys. **75**(2003)559.
[2] Brans, C.H. and Dicke, R.H.: Phys. Rev. **D124**(1961)925; Gasperini, M. and Veneziano, G.: Phys. Lett. **B277**(1992)256; Sotiriou, T.P. and Barausse, E.: Phys. Rev. **D75**(2007)084007.
[3] Buchdahl, H.A.: Mon. Not. Roy. Astron. Soc. **150**(1970)1.
[4] Kerner, R: Gen. Relativ. Gravit. **14**(1982)453; Doruisseau, J.P., Kerner, R. and Eysseric, P.: Gen. Relativ. Gravit. **15**(1983)797.
[5] Capozziello, S.: Int. J. Mod. Phys. **D11**(2002)483; Capozziello, S., Carloni, S. and Troisi, A.: Astron. Astrophys. **625**(2003)1.
[6] Nojiri, S. and Odintsov, S.D.: Phys. Rev. **D68**(2003)123512; Phys. Lett. **B576**(2003)5; Carrol, V., Duvvuri, M. and Turner, M.: Phys. Rev. **D70**(2004)043528.
[7] Meng, X. and Wang, P.: Class. Quantum Grav. **21**(2004)951; Nojiri, S. and Odintsov, S.D.: Gen. Relativ. Gravit. **36**(2004)1765.
[8] Allemandi, G. Borowiec and Francaviglia, M.: Phys. Rev. **D70**(2004)103503; Wu, X. and Zhu, Z.: Phys. Lett. **B660**(2008)293.

[9] Amendola, L. and Tsujikawa, S.: Phys. Lett. **B660**(2008)125; Cognola, G., Elizalde, E., Zerbini, S., Nojiri, S. and Odintsov, S.D.: JCAP **0502**(2005)10.
[10] Nojiri, S. and Odintsov, S.D.: J. Phys. Conf. Ser. **66**(2007)012005.
[11] de la Cruz-Dombrize, A., Dobado A. and Maroto, A.L.: Phys. Rev. **D80**(2009)124011.
[12] Kobayashi, T. and Maeda K.I.: Phys. Rev. **D78**(2008)064091; *ibid.* **D79**(2009)024009; Tsujikawa, S., Tamaki, T. and Tavakol, R.: JCAP **0905**(2009)020.
[13] Oppenheimer, J.R. and Snyder, H.: Phys. Rev. **56**(1939)455; Misner, C.W. and Sharp, D.: Phys. Rev. **136**(1964)B571.
[14] Sharif, M. and Rehmat, Z.: Gen. Relativ. Gravit. **42**(2010)1795; Sharif, M. and Fatima, S.: Gen. Relativ. Gravit. **43**(2011)127; Sharif, M. and Abbas, G.: J. Phys. Soc. Jpn. **80**(2011)104002; Sharif, M. and Kausar, H.R.: Int. J. Mod. Phys. **D** (to appear, 2011) Sharif, M. and Kausar, H.R.: Mod. Phys. Lett. **A25**(2010)3299; Sharif, M. and Kausar, H.R.: Astrophys. Space Sci. **331**(2011)281; ibid. **332**(2011)463; Sharif, M. and Kausar, H.R.: J. Phys. Soc. Jpn. **80**(2011)044004; Sharif, M. and Kausar, H.R.: Phys. Lett. **B697**(2011)01; Sharif, M. and Kausar, H.R.: JCAP **07**(2011)022.
[15] Akbar, M. and Cai, R-G: Phys. Lett. **B648**(2007)243;
Capozziello, S., Stabile, A. and Troisi, A.: Class. Quantum Grav. **25**(2008)085004;
Akbar, M. and Cai, R-G: Phys. Lett. **B635**(2006)07; Lobo, F.S.N. and Oliveira, M.A.: *Wormhole Geometries in $f(R)$ Modified Theories of Gravity*, arXiv:0909.5539v1.
[16] Misner, C.W. and Sharp, D.: Phys. Rev. **136**(1964)B571.
[17] Maartens, R.: *Causal Thermodynamics in Relativity*, arXiv:9609119; Herrera, L.: Int. J. Mod. Phys. **D15**(2006)2197.
[18] W Arnett, W.: Astrophysics J. **218**(1977)815.
[19] Martínez, J.: Phys. Rev. **D53**(1996)6921.

Journal of Physics: Conference Series **354** (2012) 012021 doi:10.1088/1742-6596/354/1/012021

Dark matter and alternative recipes for the missing mass

Crescenzo Tortora[1], Philippe Jetzer[1] and Nicola R. Napolitano[2]

[1]Universität Zürich, Institut für Theoretische Physik, Winterthurerstrasse 190, CH-8057, Zürich, Switzerland

[2]INAF – Osservatorio Astronomico di Capodimonte, Salita Moiariello, 16, 80131 - Napoli, Italy

E-mail: ctortora@physik.uzh.ch

Abstract. Within the standard cosmological scenario the Universe is found to be filled by obscure components (dark matter and dark energy) for $\sim$ 95% of its energy budget. In particular, almost all the matter content in the Universe is given by dark matter, which dominates the mass budget and drives the dynamics of galaxies and clusters of galaxies. Unfortunately, dark matter and dark energy have not been detected and no direct or indirected observations have allowed to prove their existence and amount. For this reason, some authors have suggested that a modification of Einstein Relativity or the change of the Newton's dynamics law (within a relativistic and classical framework, respectively) could allow to replace these unobserved components. We will start discussing the role of dark matter in the early-type galaxies, mainly in their central regions, investigating how its content changes as a function of the mass and the size of each galaxy and few considerations about the stellar Initial mass function have been made. In the second part of the paper we have described, as examples, some ways to overcome the dark matter hypothesis, by fitting to the observations the modified dynamics coming out from general relativistic extended theories and the MOdyfied Newtonian dynamics (MOND).

1. Introduction

Within the standard cosmological framework, the so called Λ cold dark matter (ΛCDM) scenario, the Universe is predicted to be filled by a huge amount ($\sim$ 95%) of obscure energy components (dark matter, DM, and dark energy, DE), and only $\sim$ 5% is made by baryons (i.e. stars and gas). According to several observations (i.e. distance moduli of high-z Supernovae Ia, anisotropies in the cosmic microwave background radiation, etc.) the Universe is spatially flat and in an accelerated phase of expansion. DE amounts to $\sim$ 75% of the total energy budget, is modelled as a negative equation of state and is the main driver of the observed cosmic acceleration [1, 2, 3]. DM has an important role not only at the cosmic scale but also to explain the growth of the structures and the evolution of galaxies across the cosmic time. Although this is the widest accepted theory to explain the various astronomical and cosmological observations, there has been no direct observational evidence for both DM and DE. Anyway, this has motivated the investigation of alternative theories to explain observations without invoking the presence of such unknown ingredients. We will first concentrate on DM in galaxies (mainly massive ellipticals) and on some alternative recipe. In particular we will briefly discuss the viability of two approaches: 1) a relativistic one consisting in a generalization of the Einstein' equations, the so called $f(R)$ theories, and 2) a classical one, by means of a modification of the second Newton's law, as the MOdified Newtonian Dynamics (MOND).

2. The missing mass

The need of DM in galaxies and clusters was firstly suggested by Fritz Zwicky in 1934 to account for evidence of "missing mass" in the observed orbital velocities of galaxies in clusters [4]. Applying the virial theorem to the Coma cluster he found a total mass 400 times larger than that visually observed. Till 60s, 70s no other observational evidences for the existence of DM were found. However, it was only later, in 70s, that the evidences started to be clear enough to put the DM at the center of the astronomical community attention. In these years, some authors observed that the stars and gas in spiral galaxies have an orbital velocity which remains constant even beyond their optical radii (i.e., beyond the region where much of the stars are located) [5, 6]. Nowadays, many observations have pointed out the existence of this missing mass problem in many spirals and other kinds of galaxies, as low surface brightness and ellipticals, and clusters [7, 8, 9]. The mass distribution within these different astrophysical objects is investigated by means of different mass probes. In particular, rather than circular velocity in spirals, strong (and weak) gravitational lensing, velocity dispersion measurements and X-ray observations are powerful mass tracers inside ellipticals and clusters of galaxies.

Lately, important progresses have been made also in the study of the mass distribution in elliptical galaxies, mainly thanks to the development of the discrete velocity mapping with globular clusters (GCs; [10]) and in particular the doppler measurement of the OIII emission from Planetary Nebulae [11, 12, 13, 14, 15, 16]. PNe in particular (being stars) have allowed the stellar kinematics in the (otherwise inaccessible) outer regions of such galaxies to be studied [17].

3. Dark matter

Within the Einstein framework, the missing mass needs to be filled by DM, which seems to be dominant in the very external regions of galaxies, where the stars are absent, and also in the central regions, depending on the stellar ingredients which have been adopted. Although in the central regions many uncertainties in the DM content determination arise from the contribution of stars, many kinds of observations probing the mass in these regions and for huge samples of galaxies are available. In Tortora et al. 2009 [19] and Napolitano, Romanowsky & Tortora 2010 [20], we have recently investigated the DM content within 1 $R_{\rm eff}$ (i.e., the radius where is enclosed one-half of the total projected light in the galaxy) for a sample of local ellipticals from the sample in Prugniel & Simien 1996 [18]. The stellar mass, M_*, is determined by fitting synthetic spectral models to the measured colours, while dynamical (total) mass, $M_{\rm dyn}$ within 1 $R_{\rm eff}$, is recovered by adopting a singular isothermal sphere (SIS) for the total mass distribution[1] and the observed central velocity dispersion is used to infer the best mass profile (see [19] for further details).

Djorgovski & Davis 1987 [21] and Dressler et al. 1987 [22] discovered a tight relation among central velocity dispersion σ_0, the effective radius $R_{\rm eff}$, and effective surface brightness $I_{\rm eff}$, the so-called Fundamental Plane (FP). It has been observed a deviation of its coefficients from the virial theorem expectation under the assumptions of homology and constant mass-to-light ratio (M/L). One possible explanation of the tilt is a variation of the total M/L with galaxy luminosity/mass [22], often parameterized as a power law, $M/L \propto L^{\gamma}$. This relation might reflect a variation of different galaxy properties with mass. Thus, we have evaluated the impact on γ of various factors, like stellar population properties (metallicity, age and SF history), IMF,

[1] An isothermal profile has been shown to reproduce quite well the total mass profile in massive ellipticals [28, 29]

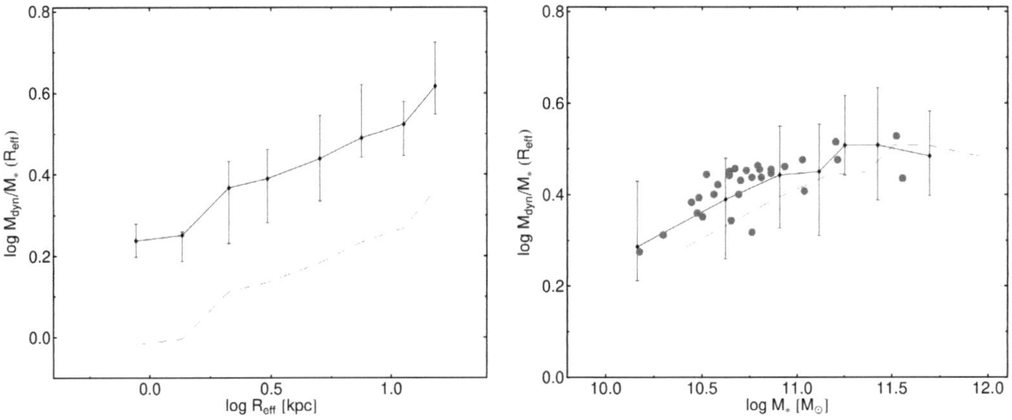

Figure 1. Dynamical-to-stellar mass ratio as a function of $R_{\rm eff}$ (left) and M_* (right) for the pure ellipticals from Prugniel & Simien 1996 [18]. Continue lines and bars are for medians and 25-75th quantiles when a Chabrier IMF is adopted, while the dashed lines are the median trends for a Salpeter IMF. The red points in the right plot are the results for simulated central galaxies in Ruszkowski & Springel 2009 [34].

rotational support, luminosity profile, non-homology and DM fraction. We found that the stellar M/L contributes little to the tilt, suggesting that the tilt is ascribed to DM.

We surveyed trends of the DM fraction within 1 $R_{\rm eff}$, finding it to be increasing with luminosity and stellar mass. Our conclusions are sensitive to various systematic uncertainties which we have investigated in detail, but are consistent with the results of dynamics studies at larger radii [19]. We also find that stronger correlations are found in terms of central velocity dispersion and effective radii, which means that at larger size, we go deeper in the halo, finding more DM dominated regions [20]. In particular, in Fig. 1 we show the deprojected dynamical-to-stellar ratio (adopting a Chabrier IMF [33]) as a function of $R_{\rm eff}$ and M_* for the pure elliptical galaxies in our sample. We find that $M_{\rm dyn}/M_* \propto {R_{\rm eff}}^{0.27}$ and $M_{\rm dyn}/M_* \propto {M_*}^{0.13}$, this last being steeper than the slope of ~ 0.06 found in Hyde & Bernardi 2009 [26]. In the smallest and less massive galaxies is $M_{\rm dyn}/M_* \sim 1.5$, i.e. DM is $\sim 40\%$ of the total mass, while in the largest and most massive galaxies is $M_{\rm dyn}/M_* \sim 3-4$, corresponding to a DM fraction of $\sim 70\%$. A Salpeter IMF [31] produces larger stellar masses, and thus lower $M_{\rm dyn}/M_*$. A rigid shift is observed in the $M_{\rm dyn}/M_* - R_{\rm eff}$ plot, null DM fractions are found in the smaller galaxies and $\sim 50\%$ of DM in the largest systems. We also compare our results with the simulated galaxies in Ruszkowski & Springel 2009 [34], finding a very good agreement in the average values, but the scatter in the simulated data is smaller. We have confirmed the increasing trend with mass and $R_{\rm eff}$ adopting a sample of intermediate redshift lens galaxies from SLACS survey [30] with available lensing and dynamical information, varying the stellar recipe and the galaxy model in [23, 24, 25].

In [20] we have also examined the correlations between masses, sizes and SF histories. We confirmed an anticorrelation between $R_{\rm eff}$ and stellar age and went on to survey for trends with the central content of DM. An average relation between the central DM density and galaxy size of $< \rho_{\rm DM} > \propto {R_{\rm eff}}^{-2}$ provides the first clear indication of cuspy DM haloes in these galaxies. From the comparison with ΛCDM expectations (see details in [20, 27]), our data are found to be consistent with a Chabrier or Kroupa IMF if the halo is adiabatically contracted[2], while,

[2] Within ΛCDM, (DM only) N-body simulations predict that galaxy and galaxy cluster halos follow a universal

if we want to retain the original outcomes from N-body simulations (i.e. the NFW profile), a Salpeter IMF is best suited (in agreement with [37]). The DM density scales with galaxy mass as expected, deviating from a universal halo profile for dwarf and spiral galaxies (see also [24, 25]).

The IMF inferred from star counts in local spiral galaxies is bottom-light (as a Chabrier or Kroupa [32]), but no direct observations are possible for other galaxy types (as, e.g., ellipticals), different environments and at higher redshift. Therefore, the IMF is a fundamental ingredient which has still to be fully understood. Anyway, Chabrier and Salpeter IMFs are assumed as the limiting cases, producing a difference in stellar masses of 0.25 dex and we have shown, that also assuming a Salpeter IMF a residual DM fraction is surviving. In order to have, on average, $M_{\rm dyn} = M_*$ we need a IMF producing stellar M/L ratios $\sim 2.7^{+1.4}_{-0.9}$ ($1\,\sigma$ uncertainty) times the ones from a Chabrier IMF. For only $\sim$ 20% of the galaxies in our sample a Salpeter IMF is totally filling the gap, avoiding the need of DM (mainly in the smallest galaxies, see left panel in Fig. 1), but for many of the galaxies larger stellar M/L would be needed. Such very high stellar masses could be produced by 1) a power-law IMF with a slope steeper than the Salpeter one[3] [38] and equivalently 2) a "bottom-heavier" and/or a "top-lighter" IMF. But, no indication exists in favour of such IMF alternatives in the local universe and they seem quite unrealistic, suggesting that DM is still an unavoidable ingredient. Anyway, recently van Dokkum & Conroy [39], analyzing the spectra of massive ellipticals in the Virgo and Coma clusters, have found that low mass stars are very abundant (contributing to $\sim$ 60% of the total stellar mass) in such galaxies and point to a steeper IMF, which in according to the comments above, would fill the gap, without needing DM, at least in the central regions.

Finally, we introduced a new fundamental constraint on galaxy formation by finding that the central DM fraction decreases with stellar age. This result is only partially explained by the size-age dependencies, and the residual trend goes on the opposite direction to basic DM halo expectations. Therefore, we suggested that there may be a connection between age and halo contraction or IMF and that galaxies forming earlier had stronger baryonic feedback, which expanded their haloes, or lumpier baryonic accretion, which avoided halo contraction, or a lighter IMF. Using the intermediate-redshift galaxy sample from the SLACS survey, we have a confirmation of such results and a negligible evidence of galaxy evolution over the last $\sim$ 2.5 Gyr other than passive stellar aging [27].

4. MOdified Newtonian gravity

The results discussed above have been obtained assuming that the Einstein relativistic theory, and the classical Newtonian theory of gravity as its limit, may be used also on galactic scales. However, the outer regions of galaxies typically are in a low acceleration regime, and in this regime Newtonian dynamics has never been experimentally tested. Motivated by this consideration, Milgrom (1983) [40] proposed to modify Newton's second law of dynamics as $F = mg$, where the acceleration g is now related to the Newtonian one $g_{\rm N}$ as $g\,\mu(g/a_0) = g_{\rm N}$. The theory thus obtained is referred to as MOND. Here, a_0 is a new universal constant and $\mu(x)$ may be an arbitrary function with the properties $\mu(x >> 1) = 1$ and $\mu(x << 1) = x$, i.e. Newton's law is recovered in high acceleration regimes, while at extremely low accelerations we have the deep-MOND regime, i.e. $F \propto g^2$. Since at large distance from the center is $\mu(g/a_0) = g/a_0$, the relation $m\mu(g/a_0)g = GMm/r^2$ becomes $g^2/a_0 = GM/r^2$ or $g = \sqrt{GMa_0}/r$. Using the relation between acceleration and velocity in a circular orbit we find $v = \sqrt[4]{GMa_0}$. Thus, MOND reproduces observed rotation curves in spiral galaxies (e.g. [41, 42]), gives a theoretical

mass density profile, the so called Navarro, Frenk & White (NFW) profile, [35]. While gas is falling down in the DM potential well, stars form and drag in the centers some DM, adiabatically contracting the DM distribution [36].

[3] In Tortora et al. 2009 [19] we have shown that if a power-law IMF with a slope of $x = -1.85$ is adopted, then the stellar M/L ratio are ~ 3.2 times the ones from a Chabrier IMF.

interpretation of the empirically determined Tully-Fisher law [41], and also works in dwarf spheroidals [43]. On the contrary, clusters seem to need extra DM to accommodate MOND [44], although recently it has been shown that sterile neutrinos can fill the gap [45].

We have considered a sample of $\sim$ 9000 local ellipticals from the SDSS sample [46] and a Chabrier IMF is adopted. For the first time we have shown that MOND is able to reproduce the dynamics in the central regions (typically $\lesssim R_{\rm eff}$) of a large sample of local ellipticals. The boost to the velocity dispersion in the MOND scenario helps to reduce the need of both radial anisotropy (in the Newtonian case without DM), deviations from a Chabrier IMF or DM. We also showed that MOND is able to predict a FP for ellipticals (similarly to the Newtonian case), but a tilt between the observed and the MOND FP is found [47]. We are working to study if MOND is able to reproduce the dynamics of both ellipticals and spirals, investigating more carefully the role of anisotropy and of the IMF.

5. Extended theories of gravity

An alternative approach to the missing mass problem consists to modify the Einstein's equations, 1) or by introducing a field (with non zero mass) which is coupled with matter, as in the so called tensor-vector-scalar (TeVeS) theories [48, 49, 50], or 2) by replacing the Ricci tensor R in the Einstein-Hilbert action with a function of R, $f(R)$. Although the TeVeS theories have been shown to be successful to reproduce different observations (e.g. [49]), we will concentrate on the $f(R)$ theories, which have been suggested to reproduce both DE on cosmic scales [51] and DM [52, 53] (see also [54, 55, 56] for a complete review on the subject).

It has been demonstrated that very general $f(R)$ analytical theories can induce a modification in the dynamics of massive particles. In particular, if one solves the field equations in the weak field limit under the general assumption of an analytic Taylor expandable $f(R)$ functions of the form

$$f(R) = f_0 + f_1 R + f_2 R^2 + ... \tag{1}$$

the new gravitational potential can be written as

$$\phi(r) = -\frac{GM}{(1+\delta)r}\left(1 + \delta e^{-\frac{r}{L}}\right) \tag{2}$$

where the first term is the Newtonian-like part of the potential associated to baryonic point-like mass $M/(1+\delta)$ (no DM) and the second term, the Yukawa-like potential, is a modification of the gravity including a scale length, L, associated to the above coefficients of the Taylor expansion. This gravitational potential was also adopted by Sanders 1984 [57], under the assumption of anti-gravity generated by massive particles (the so called Finite Length-scale Anti-Gravity, FLAG, theory). This theory conjectures that in addition to the massless graviton, particles with mass can carry the gravitational force, an anti-gravity force (if $-1 < \delta < 0$), which can allow to solve the missing mass problem. We have performed the first analysis of extended stellar kinematics (up to 7 $R_{\rm eff}$s) of 3 ellipticals (NGC 3379, NGC 4374, NGC 4494), where the Yukawa-like gravitational potential in Eq. (2) is considered [58]. We find that these modified potentials are able to fit quite well all galaxies in our sample and the orbital anisotropy distribution turns out to be similar to the one estimated if a dark halo is considered. The parameter which measures the strength of the Yukawa-like correction, δ ($= -0.88, -0.79, -0.75$ for the three galaxies) is, on average, larger than the one found previously in spiral galaxies ([57], $-0.95 \lesssim \delta \lesssim -0.92$) and seemingly correlating with the orbital anisotropy.

Though the additional Yukawa term in the gravitational potential modifies dynamics with respect to the standard Newtonian limit of General Relativity, we have demonstrated that the motion of massless particles results unaffected thanks to suitable cancellations in the post-Newtonian limit [59]. Thus, all the lensing observables are equal to the ones known from General

Relativity. We are planning to test the implications of these results by fitting both lensing and dynamical observables in galaxy and cluster lensing events. In particular, following [27], the sample of lens galaxies at intermediate redshift from SLACS sample will be an efficient testing ground for these extended theories.

6. Conclusions

In this contribution we have discussed the missing mass problem, which come out from different observations in galaxies and clusters. In a standard Newtonian and Einstein approach, DM is an unavoidable ingredients to explain both cosmological observations and dynamics in galaxies and clusters. DM is found to be the dominant component, at all, since the stellar content is negligible with respect to the total mass determined by, e.g. dynamical analysis. Less obvious is the DM role and its amount in the central regions (typically $\lesssim R_{\rm eff}$) where a very large fraction of stars are settled. We have shown that DM in the central regions is strongly dependent on the IMF assumption. In fact, while the typical "bottom-light" (Chabrier or Kroupa) IMFs, found through direct observations in local spirals, predict lower stellar masses and thus larger DM fractions, $\sim 40-70\%$ in massive ellipticals (with stellar mass $M_* \gtrsim 10^{10} M_{\odot}$), a Salpeter IMF predict larger stellar masses, and thus lower DM fractions $\sim 0-50\%$. We have also shown how the typical halo density profiles derived from ΛCDM N-body simulations are consistent with the only if stars are distributed in according to a Salpeter IMF, while a contraction of the DM halo is needed to reconcile a Chabrier IMF. Moreover, to avoid any DM in the central regions, we would need steeper IMFs, but no strong observation seems now convincing us which this is the case (e.g. [39]). The situation in the external regions (few $R_{\rm eff}$s up to the virial radius) is more complicated, and such a steep IMF seems to us not enough to replace the amount of missing mass needing in such external regions.

Anyway, until no direct indication of the existence of DM will be provided by the many working (or future) experiments, people are encouraged to search for some alternative theories to overcome the need of such unseen matter source. This can be done by means of different approaches, within both a relativistic theory, as an extension of the General Relativity, and within a classical Newtonian approach, like in the MOND. We have shown as both MOND or $f(R)$ theories are able to reproduce a wide set of observations, in particular, the dynamics in different kinds of galaxies, and recently both central ($\lesssim R_{\rm eff}$) and more external (few $R_{\rm eff}$) dynamical data in ellipticals [47, 58].

In the future we expect to investigate the correlations between DM content, mass, IMF and stellar population parameters, checking what is the impact on such correlations when the gravity theory is changed. Such theories need to be tested on wider samples of (different type) galaxies, and galaxy clusters, relying on different kinds of observations (gravitational lensing, dynamics, X-ray, etc.), in order to check if DM can be avoided at all, everywhere in galaxies or if our constrain on DM in galaxies have to be revised downward. To refuse/validate any of such theories, as alternatives to DM, we need a) to fit data quite good, at least as much as in the DM case [58], b) that some parameters derived from the fitting procedure (as other model parameters, velocity dispersion anisotropy, stellar mass-to-light ratio if allowed to be free, etc.) assume physically reasonable values [47, 58] and c) different observational probes give consistent results. If such points would be addressed, then the theory analyzed can be considered as a viable alternative. Anyway, a definite answer about the existence of DM and its abundance will arrive from both direct and indirect experiments in progress and scheduled to start in the few next years.

References

[1] Perlmutter S. et al. 1999 *ApJ* **517** 565
[2] Reiss A.G. et al. 1998 *AJ* **116** 1009

[3] Spergel D.N. et al. 2007 *ApJS* **170** 377
[4] Zwicky, F. 1937 *ApJ* **86** 217
[5] Rubin V. C. & Ford W. K. Jr. 1970 *ApJ* **159** 37
[6] Bosma A. 1981 *AJ* **86** 1825
[7] de Blok W. J. G., McGaugh S. S., Bosma A. & Rubin V. C. 2001 *ApJ* **552** 23
[8] Salucci P. & Borriello A. 2003 *Particle Physics in the New Millennium* **616** 66
[9] Bradač M. et al. 2008 *ApJ* **687** 959
[10] Romanowsky A. J., Strader J., Spitler L. R., Johnson R., Brodie J. P., Forbes D. A. & Ponman T. 2009 *AJ* **137** 4956
[11] Napolitano, N. R., Arnaboldi, M., Freeman, K. C., & Capaccioli, M. 2001 *A&A* **377** 784
[12] Napolitano, N. R., Arnaboldi, M., & Capaccioli, M. 2002, *A&A* **383** 791
[13] Romanowsky A. J., Douglas N. G., Arnaboldi M., Kuijken K., Merrifield M. R., Napolitano N. R., Capaccioli M. & Freeman, K. C. 2003 *Science* **301** 1696
[14] De Lorenzi F. et al. 2009 *MNRAS* **395** 76
[15] Napolitano N. R. et al. 2009 *MNRAS* **393** 329
[16] Napolitano N. R. et al. 2011 *MNRAS* **411** 2035
[17] Coccato L. et al. 2009 *MNRAS* **394** 1249
[18] Prugniel Ph. & Simien F. 1996 *A&A* **309** 749
[19] Tortora C. et al., 2009 *MNRAS* **396** 1132
[20] Napolitano N. R., Romanowsky A.J. & Tortora C., 2010 *MNRAS* **405** 2351
[21] Djorgovski S. & Davis M. 1987 *ApJ* **313** 59
[22] Dressler A., Lynden-Bell D., Burstein D., Davies R. L., Faber S. M., Terlevich R., Wegner G., 1987 *ApJ* **313** 42
[23] Cardone V. F., Tortora C., Molinaro R., Salzano V., 2009 *A&A* **504** 769
[24] Cardone V.F. & Tortora C., 2010 *MNRAS* **409** 1570
[25] Cardone V. F., Del Popolo A., Tortora C., Napolitano N.R. 2011 *MNRAS* **416** 1822
[26] Hyde J. B. & Bernardi M. 2009 *MNRAS* **394** 1978
[27] Tortora C., Napolitano N.R., Romanowsky A.J., Jetzer Ph. 2010 *ApJ* **721** 1
[28] Koopmans L. V. E., Treu T., Bolton A. S., Burles S., Moustakas L. A. 2006 *ApJ* **649** 599
[29] Gavazzi R. et al. 2007 *ApJ* **667** 176
[30] Auger M. W., Treu T., Bolton A. S., Gavazzi R., Koopmans L. V. E., Marshall P. J., Bundy K. & Moustakas L. A. 2009 *ApJ* **705** 1099
[31] Salpeter E.E. 1955 *ApJ* **121** 161
[32] Kroupa P. 2001 *MNRAS* **322** 231
[33] Chabrier G. 2001 *ApJ* **554** 1274
[34] Ruszkowski M. & Springel V. 2009, ApJ, 696, 1094
[35] Navarro J. F., Frenk C.F. & White S.D.M. 1996 *ApJ* **462** 563
[36] Gnedin O. Y., Kravtsov A. V., Klypin A. A. & Nagai D. 2004 *ApJ* **616** 16
[37] Treu T., AugerM. W., Koopmans L. V. E., Gavazzi R., Marsh'all P. J., Bolton A. S. 2010 *ApJ* **709** 1195
[38] Bell E. F. & de Jong R. S. 2001 *ApJ* **550** 212
[39] van Dokkum P. G. & Conroy C. 2010 *Nature* **468** 940
[40] Milgrom M., 1983 *ApJ* **270** 365
[41] McGaugh S. S., 2005 *ApJ* **632** 859
[42] McGaugh S. S., 2008 *ApJ* **683** 137
[43] Angus G. W. 2008 *MNRAS* **387** 1481
[44] Pointecouteau E. and Silk J. 2005 *MNRAS* **364** 654
[45] Angus G. W. 2009 *MNRAS* **394** 527
[46] Blanton M. R., Lupton R. H., Schlegel D. J., Strauss M. A., Brinkmann J., Fukugita M., Loveday J. 2005 *ApJ* **631** 208
[47] Cardone V.F., Angus G., Diaferio A., Tortora C., Molinaro R. 2011 *MNRAS* **412** 2617
[48] Moffat J. W. 2006 *JCAP* **03** 004
[49] Brownstein J. R., Moffat J. W. 2007 *MNRAS* **382** 29
[50] Moffat J. W., Toth V. T. 2009 *MNRAS* **397** 1885
[51] Capozziello S., Cardone V. F., Carloni S. and Troisi A., 2003 *International Journal of Modern Physics* **D12** 1969.
[52] Sanders R.H. and McGaugh S. S., 2002 *ARA&A* **40** 263
[53] Capozziello S., Piedipalumbo E., Rubano C. and Scudellaro P., 2009 *A&A* **505** 21
[54] Capozziello S., de Laurentis M. 2011 *PhR* **509** 167
[55] Clifton T. Ferreira, Pedro G., Padilla A., Skordis C. 2011, arXiv:1106.2476

[56] Nojiri S., Odintsov S. D. 2011 *PhR* **505** 59
[57] Sanders R. H. 1984 *A&A* **136** L21
[58] Napolitano N.R., Capozziello S., Romanowsky A.J., Capaccioli M. & Tortora C. 2012, to appear in ApJ, arXiv:1201.3363
[59] Lubini, M., Tortora, C., Näf, J., Jetzer, Ph., Capozziello, S. 2011, *EPJC* **71** 1834, arXiv:1104.2851

AUTHOR INDEX